SCIENCE *of*

UNDERSTAND THE ANATOMY AND PHYSIOLOGY TO PERFECT YOUR PRACTICE

YOGA

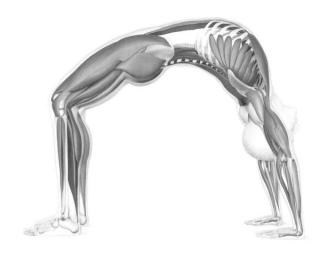

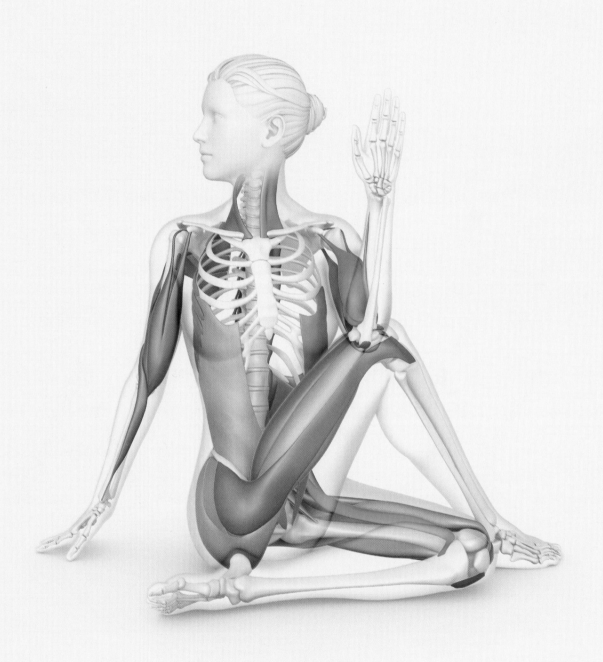

SCIENCE *of*
UNDERSTAND THE ANATOMY AND PHYSIOLOGY TO PERFECT YOUR PRACTICE
YOGA

Ann Swanson

Senior Editor **Ruth O'Rourke-Jones**
Senior Designer **Clare Joyce**
Editor **Alice Horne**
Senior Art Editor **Karen Constanti**
Design assistance **Philippa Nash,
Louise Brigenshaw**
Editorial Assistant **Megan Lea**
Senior Jacket Creative **Nicola Powling**
Jacket Co-ordinator **Lucy Philpott**
Pre-production Producers
Heather Blagden, Tony Phipps
Production Controller **Igrain Roberts**
Creative Technical Support
Sonia Charbonnier
Managing Editor **Dawn Henderson**
Managing Art Editor
Marianne Markham
Art Director **Maxine Pedliham**
Publishing Director
Mary-Clare Jerram

Illustrations **Arran Lewis**

First published in Great Britain in 2019 by
Dorling Kindersley Limited,
80 Strand, London WC2R 0RL
Copyright © 2019
Dorling Kindersley Limited
A Penguin Random House Company
2 4 6 8 10 9 7 5 3 1
001–310296–Jan/19
All rights reserved.
No part of this publication may be reproduced,
stored in or introduced into a retrieval system,
or transmitted, in any form, or by any means
(electronic, mechanical, photocopying,
recording, or otherwise) without the prior
written permission of the copyright owners.

A CIP catalogue record for this book
is available from the British Library.
ISBN: 978-0-2413-4123-0
Printed and bound in China

A WORLD OF IDEAS:
SEE ALL THERE IS TO KNOW
www.dk.com

CONTENTS

HUMAN ANATOMY 8

THE ASANAS 42

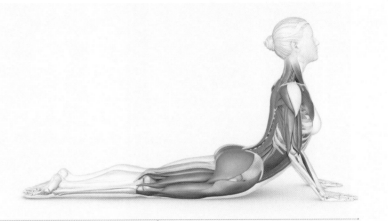

FOREWORD

As the daughter of a NASA scientist I was raised to have an analytical mind. A part of me craves method, data, and evidence. I started journaling aged seven, carrying my notebooks everywhere. I filled them with charts, graphs, observations, and plans concerning everything from what I ate that day to what to rent at the video store.

I was a curious child, constantly asking "Why?". My parents would send me to the trusty encyclopedia to look up the answer.

At the same time, I have always been artistic, creative, and interested in spirituality. My notebooks are also filled with elaborate stories, poetry, and colourful drawings.

My undergraduate studies in art led to burn-out. Like many people, I came to yoga hoping to relieve stress and anxiety during a difficult time – with the added bonus of staying fit. I didn't expect that yoga would transform me in an ineffable, seemingly magical way.

When I started practising, I aimed to make the picture-perfect poses. I slowly realized that yoga isn't about performing the pose "perfectly", but instead about being perfectly okay with my body and mind in the moment. Now I know that many of the most profound effects of poses transcend my anatomy of muscles and bones to shape my neurology, psychology, and energetic body.

I vividly remember lying on my mat at the end of a yoga class with my eyes wide open looking impatiently around when I was supposed to be relaxing. I thought "What a waste of time; I have work to do!" With practice, I started to enjoy the way relaxation and meditation practices made me feel.

Now, through reading research, I know that when I meditate, I am literally reshaping my brain. Ultimately, I am impacting every single system of my body, optimizing function. What more important work could I possibly do?

My shifting mindset drew me to the Himalayas to study yoga, massage, and healing arts. My teacher, Yogi Sivadas, renewed my interest in science. I returned to the US and completed the pre-medicine courses, in pursuit of understanding how and why yoga works in such life-changing ways.

I will never forget the first time I held a human brain in the cadaver lab. The experience was neither antiseptic nor clinical, but deeply spiritual. That three-pound folded grey mysterious mass once both computed mathematics and felt the depths of love. Holding that brain, I knew that the mind-body connection was a key mechanism behind yoga's benefits.

> **Scientific** principles and evidence have **demystified** so much of the practice

Science of Yoga is the book I wanted to read when I first started practising yoga. In classes, teachers offer (sometimes conflicting) cues and claims – "Calm your nervous system by elongating your exhales", "This pose will boost your immunity", "Align your knee over your ankle" – and I constantly wondered "Why?".

For the past decade, through workshops, reading research papers, and completing my Master of Science in yoga therapy at Maryland University of Integrative Health, I have continued to fill my notebooks with facts, figures, sketches, and stories. *Science of Yoga* summarizes the notes I found most fascinating as a yoga student and teacher. This book is intended as neither a comprehensive text on human anatomy and yoga, nor a medical reference book; it is just the beginning. My intention is for this material to spark more curiosity and discussion about the science of yoga, leading to more inspired yoga practitioners and professionals, more rigorous research, more public policies that encourage yoga in schools and healthcare, and, ultimately, more accessibility and acceptance.

Through my research, scientific principles and evidence have demystified so much of the practice. Surprisingly, this made my transformative experiences feel even more magical. There is just so much more to discover. In the grand scheme of scientific enquiry, yoga research is in its infancy. However, now is an exciting and pivotal point in the field, with a remarkable increase in the quality and quantity of yoga research papers in the past few decades; the evidence supporting yoga's benefits continues to grow rapidly.

Science can explain the hows and whys of many things, but research studies, no matter how rigorously conducted, cannot compare to your personal, experiential evidence of healing and transformation. Only you can harness the power of yoga through practice. As with any scientific inquiry, I hope this book leaves you with more questions than answers, bringing out your inner child to playfully enquire "Why?".

Be well,

Ann Swanson
Mind-body science educator and certified yoga therapist
www.AnnSwansonWellness.com

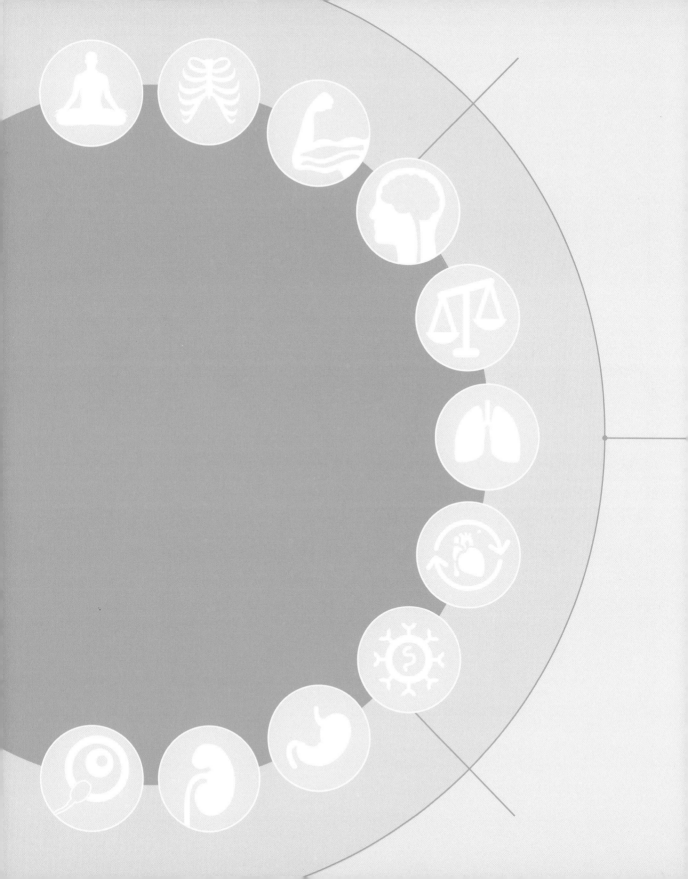

HUMAN ANATOMY

Most yoga anatomy books and courses focus on the musculoskeletal system, but research shows that practising yoga affects all body systems. This section breaks down the key effects and benefits for each one. Study your anatomical systems as modern biology defines them – then, challenge yourself to shift to a yogic perspective, one of unity. Experience your extraordinary body as an interconnected whole.

CELL TO SYSTEM

As in design, a key concept in biology is "form follows function" – this means that the physical structures of your body reflect their specific tasks. Anatomy is the study of these body structures and physiology is the study of their functions, or how your body works.

Telomeres

Telomeres are like caps on the tips of chromosomes. With ageing, telomeres tend to shorten. Studies on the cutting edge of molecular biology have shown that a yogic lifestyle (including asanas, meditation, social support, and a plant-based diet) seems to increase telomere length, which may have an impact on increased longevity and health.

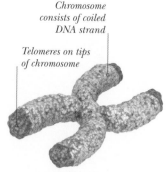

Chromosome consists of coiled DNA strand

Telomeres on tips of chromosome

CHROMOSOME

BUILDING BLOCKS

Atoms are the building blocks of matter; cells are the building blocks of biological life. Approximately 37 trillion body cells are vibrating in your body right now. They create four basic tissue types and 11 organ systems. All of these parts and pieces create an integrated whole called the human body.

Liver cells are called hepatocytes

Blood vessels

Cell membrane is semipermeable outer layer

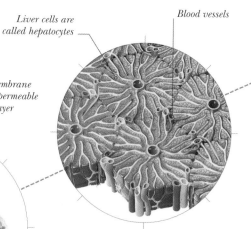

Tissue
Cells come together to form tissues, which are like unique fabrics. This specifically shaped tissue is located in the liver.

Electrons surround the nucleus

Protons and neutrons are in the nucleus

Cell
Cells are the smallest unit of life. Most cells contain a nucleus at the centre, jelly-like cytoplasm, and an outer layer called the cell membrane. Small functional units inside the cell are called organelles.

DNA contains the information a cell needs to function and replicate

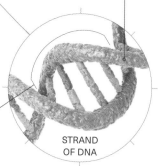

Atom
These chemical building blocks contain protons, neutrons, and electrons. They bond together to make important molecules, such as water (H_2O).

A gene is a unit of DNA in a cell nucleus – meditation may prevent cellular ageing and harmful gene expression

STRAND OF DNA

Organ

Tissues come together to form organs, like your liver (shown below). This large organ receives blood from all over your body for processing and purification. It also makes bile, which is used to break down fats in the digestive process.

Liver forms part of digestive system

Digestive system absorbs nutrients and eliminates waste products

Liver has two lobes

System

Organs come together to form organ systems, including: integumentary, skeletal, muscular, nervous, endocrine, respiratory, cardiovascular, lymphatic, digestive (shown above), urinary, and reproductive.

Integumentary system

The integumentary system includes hair, nails, skin and associated structures like sweat glands. Some claim that hot yoga causes you to "sweat out toxins". However, your liver is responsible for such detoxification processes. What you are actually sweating out is water, leading to dehydration. If you sweat a lot or practise hot yoga, make sure you drink plenty of water to replenish your losses.

Tactile nerve made of nervous tissue

Sweat gland

Dermis made of connective tissue

Epidermis made of epithelial tissue

Arrector pili made of muscle tissue

Hair

SKIN

The skin has two main layers: the epidermis at the surface and the dermis below, which contains sweat glands, blood vessels, nerves, and hair follicles.

Blood vessel

Nerve

Human body

Organ systems come together to form an organism. As a human being, you are made up of all of this, functioning as a dynamic, living whole.

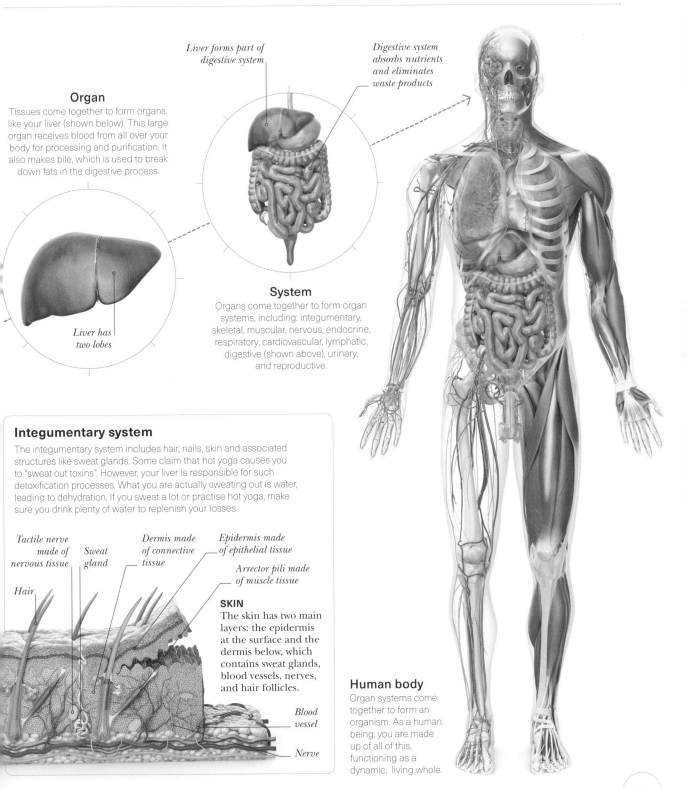

SKELETAL
SYSTEM

The 206 bones that make up your skeleton are dynamic, living organs. Together they form a framework for your body that provides structure, protection, and has the ability to move.

SYSTEM OVERVIEW

Your bones are made of collagen and they store calcium, a mineral that makes them strong and is vital for bodily functions. They also contain bone marrow where blood cells are produced. Bones form joints, which are supported by cartilage and structures such as ligaments. Yoga can support your bone and joint health.

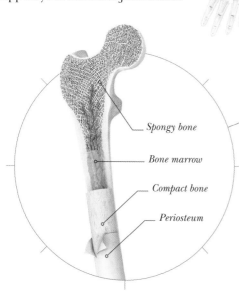

Spongy bone

Bone marrow

Compact bone

Periosteum

Bone structure

Bone has a smooth outer connective tissue shell called periosteum. Inside this is a strong, dense layer known as compact bone. Deeper still, is spongy bone with honeycomb-like spaces; this is strong yet light.

Skull
These fused plates of bone protect your brain.

Mandible
Lower jaw bone that forms the only movable joint in your skull.

Clavicle
Also called the collar bone, it connects your scapulae and sternum.

Sternum
Also called the breastbone, connects your ribs.

Ribs
The 12 pairs of bones that form your ribcage.

Pelvis
Two hip bones connected by your sacrum.

Carpals
Eight small bones form each wrist.

Metacarpals
Five long bones ru through each palm

Phalanges
Each hand has 14 bones forming your fingers.

Patella
Also called the knee cap, attached to your quadriceps tendon.

Tarsals
The seven small bones that form your ankle.

Metatarsals
Five long bones that run through your foot.

Phalanges
The 14 bones in each foot that form your toes.

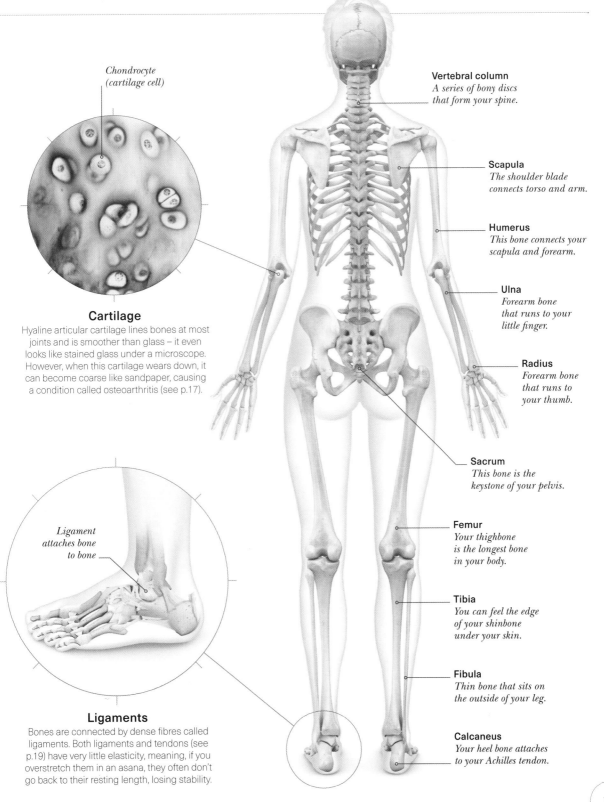

Chondrocyte
(cartilage cell)

Cartilage

Hyaline articular cartilage lines bones at most joints and is smoother than glass – it even looks like stained glass under a microscope. However, when this cartilage wears down, it can become coarse like sandpaper, causing a condition called osteoarthritis (see p.17).

Ligament
attaches bone
to bone

Ligaments

Bones are connected by dense fibres called ligaments. Both ligaments and tendons (see p.19) have very little elasticity, meaning, if you overstretch them in an asana, they often don't go back to their resting length, losing stability.

Vertebral column
A series of bony discs
that form your spine.

Scapula
The shoulder blade
connects torso and arm.

Humerus
This bone connects your
scapula and forearm.

Ulna
Forearm bone
that runs to your
little finger.

Radius
Forearm bone
that runs to
your thumb.

Sacrum
This bone is the
keystone of your pelvis.

Femur
Your thighbone
is the longest bone
in your body.

Tibia
You can feel the edge
of your shinbone
under your skin.

Fibula
Thin bone that sits on
the outside of your leg.

Calcaneus
Your heel bone attaches
to your Achilles tendon.

SPINE

Your vertebrae sit on top of each other to create natural curves. This is called a "neutral spine". It alternates between curving inward (lordosis) and outward (kyphosis) to absorb shock like a coiled spring. Your vertebrae are like wedges stacked to form these curves in order to bear your body weight most efficiently.

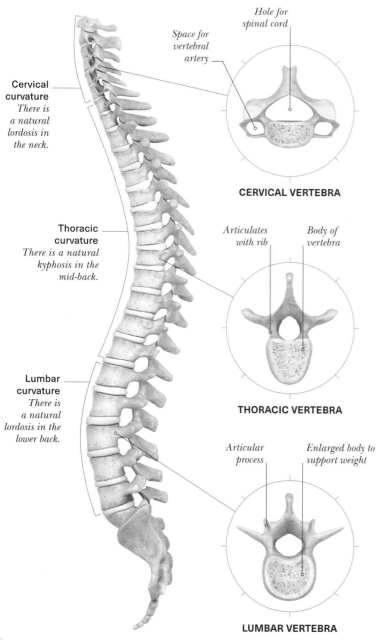

Cervical curvature
There is a natural lordosis in the neck.

Space for vertebral artery

Hole for spinal cord

CERVICAL VERTEBRA

Thoracic curvature
There is a natural kyphosis in the mid-back.

Articulates with rib

Body of vertebra

THORACIC VERTEBRA

Lumbar curvature
There is a natural lordosis in the lower back.

Articular process

Enlarged body to support weight

LUMBAR VERTEBRA

 Neutral spine

Many asanas incorporate a neutral spine, such as seated meditation poses. Poor posture and other considerations can lead to a multitude of spinal structural deviations, including common conditions like hyperlordosis and hyperkyphosis. Yoga works your spine in novel ways and enhances body awareness to improve your overall posture.

Gentle, even curves

NEUTRAL SPINE
These natural curves create the strongest, most stable alignment of the spine. In this ideal, the spine is also not twisted or leaning to either side.

Curvature in upper spinal column

KYPHOSIS
Hyperkyphosis of the thoracic spine is often simply called a kyphosis or hunchback. This exaggerated curvature is common in osteoporosis.

Curvature in lower spinal column

LORDOSIS
Hyperlordosis of the lumbar spine is sometimes just called a lordosis or swayback. This exaggerated curvature is natural during pregnancy.

PELVIS

Your pelvis includes two hip (coxal) bones connected by your sacrum. The sacrum, which means "sacred" in Latin, is the triangular bone with the tailbone at the lower, or inferior, end; it acts like the keystone to an arched bridge, forming a structurally sound base for your spine.

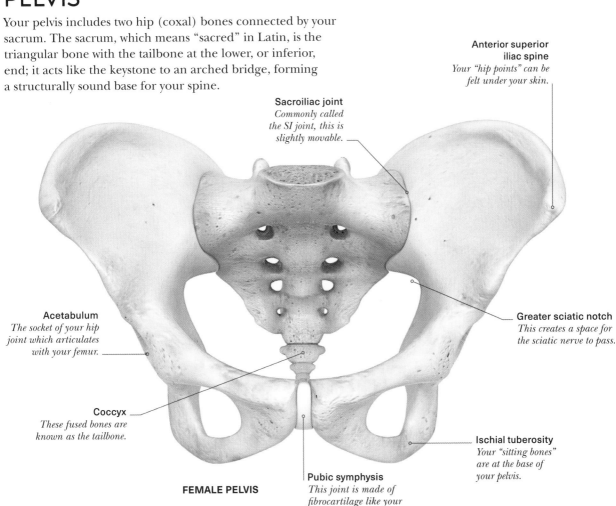

Anterior superior iliac spine
Your "hip points" can be felt under your skin.

Sacroiliac joint
Commonly called the SI joint, this is slightly movable.

Acetabulum
The socket of your hip joint which articulates with your femur.

Greater sciatic notch
This creates a space for the sciatic nerve to pass.

Coccyx
These fused bones are known as the tailbone.

Ischial tuberosity
Your "sitting bones" are at the base of your pelvis.

FEMALE PELVIS

Pubic symphysis
This joint is made of fibrocartilage like your intervertebral discs.

Neutral pelvis

A neutral pelvis facilitates a neutral spine and vice versa. Imagine your pelvic bowl filled with water. Finding a neutral spine and pelvis means that the water wouldn't spill backwards, forwards, or to the side – such as when one of your hip points are lifted or your pelvis is rotated.

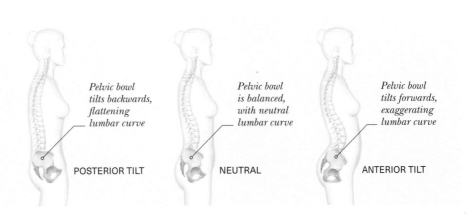

Pelvic bowl tilts backwards, flattening lumbar curve

POSTERIOR TILT

Pelvic bowl is balanced, with neutral lumbar curve

NEUTRAL

Pelvic bowl tilts forwards, exaggerating lumbar curve

ANTERIOR TILT

JOINTS

Joints are where bones unite and articulate to allow movement. There are three joint types: fibrous, cartilaginous, and synovial. Fibrous joints are immobile, such as the sutures in your skull. Cartilaginous joints are slightly mobile, like your pubic symphysis. Synovial joints are most mobile and are very important for asanas.

JOINT ACTIONS

Synovial joints of your body can move in many directions. Hinge joints in your elbow and knee mainly perform flexion and extension, like the hinge of a door. Larger ball and socket joints like in your shoulder and hip can also perform abduction, adduction, rotation, and circumduction, which is a combination of all of the above movements.

TYPES OF MOVEMENT

Flexion	Angle at joint generally gets smaller
Extension	Angle at joint generally gets larger
Abduction	A limb moves away from the body
Adduction	A limb moves closer towards the body
External rotation	A limb rotates outwards
Internal rotation	A limb rotates inwards
Axial rotation	The spine twists on its axis
Plantar flexion	Pointing the feet
Dorsiflexion	Flexing the feet

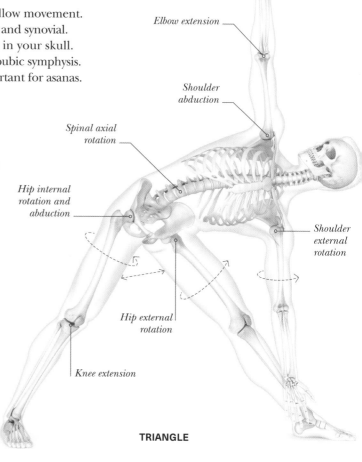

Elbow extension

Shoulder abduction

Spinal axial rotation

Hip internal rotation and abduction

Shoulder external rotation

Hip external rotation

Knee extension

TRIANGLE

Inside a joint

Synovial fluid lubricates and cushions. It is a "non-Newtonian fluid", which means it gets more viscous or thicker in response to pressure, similar to solutions of cornstarch in water. With a sedentary lifestyle, synovial fluid may become thin and less effective. However, impact from the practice of yoga asanas causes synovial fluid to thicken, better protecting joint structures such as cartilage and reducing pain.

SYNOVIAL JOINT

Synovial joints allow movement while protecting bone ends from touching each other, which would cause damage. They are the most common type of joint in the body.

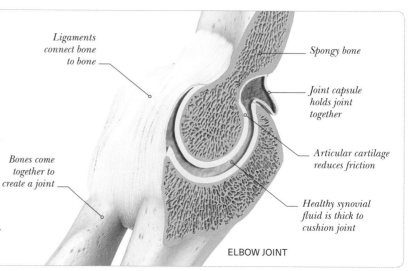

Ligaments connect bone to bone

Bones come together to create a joint

Spongy bone

Joint capsule holds joint together

Articular cartilage reduces friction

Healthy synovial fluid is thick to cushion joint

ELBOW JOINT

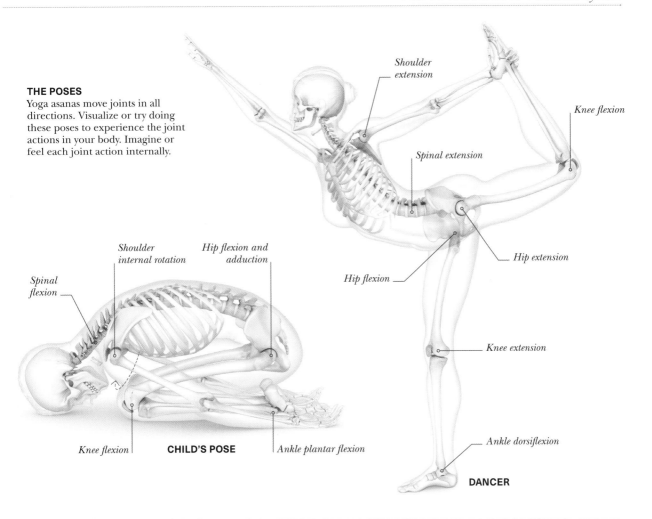

THE POSES

Yoga asanas move joints in all directions. Visualize or try doing these poses to experience the joint actions in your body. Imagine or feel each joint action internally.

Shoulder extension

Knee flexion

Spinal extension

Hip extension

Hip flexion

Spinal flexion

Shoulder internal rotation

Hip flexion and adduction

Knee extension

Knee flexion   **CHILD'S POSE**   *Ankle plantar flexion*

Ankle dorsiflexion

DANCER

Arthritis

Wear and tear on joints can lead to osteoarthritis. In a 7-year clinical trial, researchers found that yoga is both safe and effective in managing both osteoarthritis and rheumatoid arthritis (see p.37). After an 8-week yoga class, participants showed a reduction in pain by 25 per cent, along with statistically significant improvements in physical fitness and quality of life.

PROGRESSION

As cartilage degrades there is less space in the joint, leading to inflammation and pain. Bone spurs or osteophytes can form as the condition progresses.

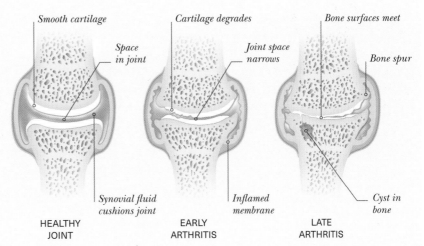

Smooth cartilage

Space in joint

Cartilage degrades

Joint space narrows

Bone surfaces meet

Bone spur

Synovial fluid cushions joint

Inflamed membrane

Cyst in bone

HEALTHY JOINT   EARLY ARTHRITIS   LATE ARTHRITIS

MUSCULAR
SYSTEM

There are around 640 muscles in your body. Your skeletal muscles are attached to your bones, allowing you to move. Some muscles are superficial (close to the surface) and others are deep.

SYSTEM OVERVIEW

As you study each of these key chosen muscles, try to palpate or physically touch them while visualizing their internal location. This will help you learn better, while improving your mind-body connection. Most of the muscles here are categorized into groups based on their actions.

Muscle fibres are arranged in parallel orientation

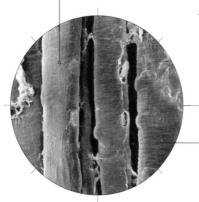

Striations are visible stripes from internal structures (see p.21)

Skeletal muscle
There are three types of muscle tissue: cardiac, smooth, and skeletal, but we will focus on skeletal muscle as it is responsible for the movement of joints in asana. This is what it looks like under a microscope.

Pectorals
Pectoralis major
Pectoralis minor

Intercostal muscles

Brachialis

Abdominals
Rectus abdominis
External abdominal obliques
Internal abdominal obliques
(deep, not shown)
Transversus abdominis

Hip flexors
Iliopsoas (iliacus and psoas major
Rectus femoris
(see quadriceps)
Sartorius
Adductors
(see below)

Adductors
Adductor longus
Adductor brevis
Adductor magnus
Pectineus
Gracilis

Quadriceps
Rectus femoris
Vastus medialis
Vastus lateralis
Vastus intermedius
(deep, not shown)

Ankle dorsiflexors
Tibialis anterior
Extensor digitorum longus
Extensor hallucis longus

Elbow flexors
Biceps brachii
Brachialis (deep)
Brachioradialis

SUPERFICIAL

DEEP

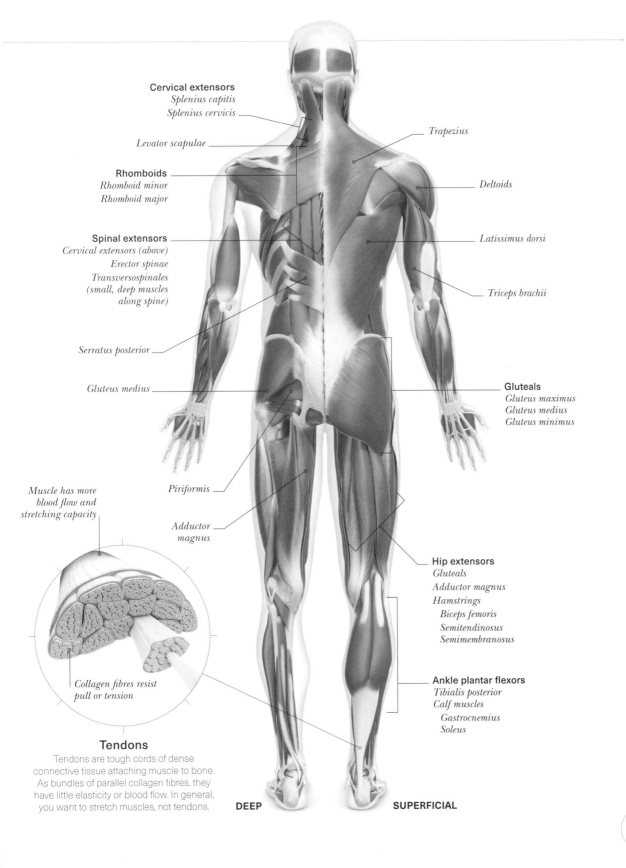

Cervical extensors
Splenius capitis
Splenius cervicis

Levator scapulae

Rhomboids
Rhomboid minor
Rhomboid major

Spinal extensors
Cervical extensors (above)
Erector spinae
Transversospinales
(small, deep muscles
along spine)

Serratus posterior

Gluteus medius

Muscle has more
blood flow and
stretching capacity

Piriformis

Adductor
magnus

Trapezius

Deltoids

Latissimus dorsi

Triceps brachii

Gluteals
Gluteus maximus
Gluteus medius
Gluteus minimus

Hip extensors
Gluteals
Adductor magnus
Hamstrings
Biceps femoris
Semitendinosus
Semimembranosus

Ankle plantar flexors
Tibialis posterior
Calf muscles
Gastrocnemius
Soleus

Collagen fibres resist
pull or tension

Tendons
Tendons are tough cords of dense
connective tissue attaching muscle to bone.
As bundles of parallel collagen fibres, they
have little elasticity or blood flow. In general,
you want to stretch muscles, not tendons.

DEEP

SUPERFICIAL

MUSCLE STRUCTURE

Skeletal muscles are bundles of bundles of bundles of parallel muscle cells, blood vessels, and nerves wrapped with connective tissue, including fascia. Fascia creates a network through and around muscles and other structures of your body. Microscopic proteins in your muscles cause muscle contractions.

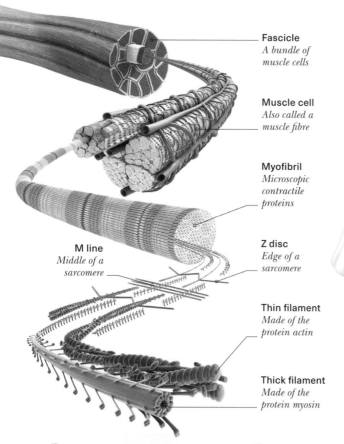

Fascicle
A bundle of muscle cells

Muscle cell
Also called a muscle fibre

Myofibril
Microscopic contractile proteins

M line
Middle of a sarcomere

Z disc
Edge of a sarcomere

Thin filament
Made of the protein actin

Thick filament
Made of the protein myosin

Fascia

Fascia is similar to the white pith of an orange; it separates parts yet integrates the whole. Fascia is not just found around muscles. It is also around vital organs and woven throughout your body. This body-wide network of fascia is part of the reason why a yoga pose that affects your foot can suddenly release your tight shoulders.

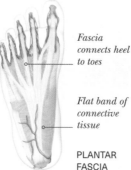

Fascia connects heel to toes

Flat band of connective tissue

PLANTAR FASCIA

HOW MUSCLES WORK

Muscles often work in antagonistic pairs. As the agonist muscle engages, the antagonist generally releases. Synergist muscles engage around the joint to support the action.

TYPES OF CONTRACTION

Isotonic contractions involve a change in muscle length, as in the act of flexing or extending your elbow (see below) or transitioning in or out of an asana. Isometric contractions involve tension with no change in muscle length, such as when holding an asana.

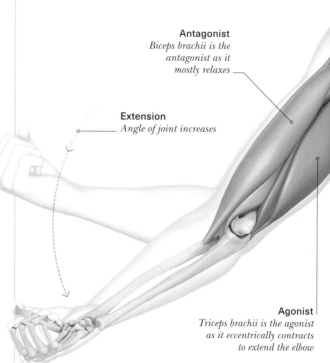

Antagonist
Biceps brachii is the antagonist as it mostly relaxes

Extension
Angle of joint increases

Agonist
Triceps brachii is the agonist as it eccentrically contracts to extend the elbow

ECCENTRIC CONTRACTION

Eccentric contractions occur when muscle fibres "lengthen" to change the angle of a joint. This occurs when extending your elbow or extending your knee as you transition from Warrior II to Triangle (see pp.118–21) pose.

Movement and fascia

Research suggests that the collagen fibres of the fascia surrounding healthy muscles are organized in a criss-cross, lattice structure. Inactivity and ageing seem to cause your fascia to lose its structural integrity. Asana may help organize your fascia, helping you move and feel better.

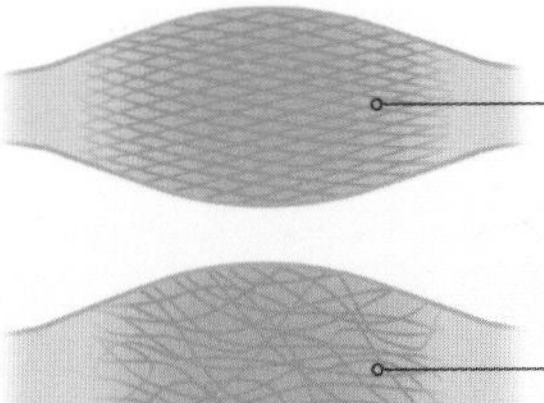

Healthy double lattice collagen orientation

Random collagen orientation from inactivity

COLLAGEN FIBRES

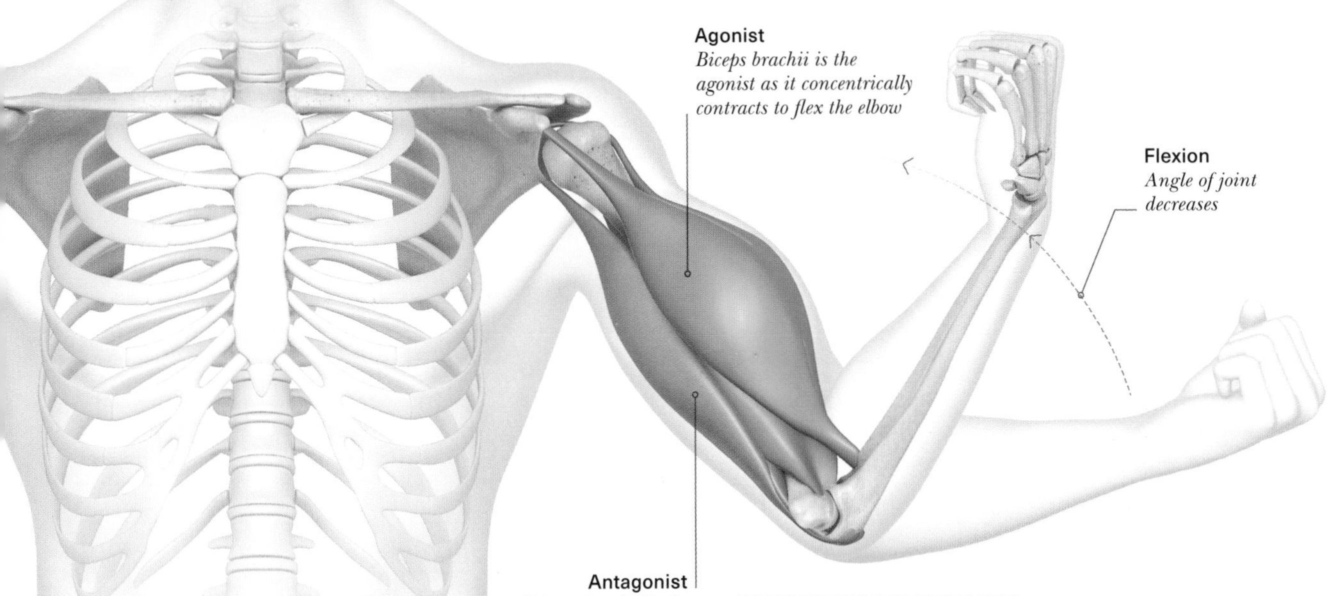

Agonist
Biceps brachii is the agonist as it concentrically contracts to flex the elbow

Flexion
Angle of joint decreases

Antagonist
Triceps brachii is the antagonist as it mostly relaxes

CONCENTRIC CONTRACTION
Concentric contractions occur when the muscle fibres "shorten" to change the angle of a joint. This occurs when flexing your elbow as you lift a weight or flexing your knee as you move into Warrior II (see pp.102-05).

Muscle contraction

A cascade of events initiated by a signal from the nervous system and the presence of calcium leads to the removal of the blockage on actin of the thin filament, allowing the thick and thin filament to connect. The thick filament pulls the thin filament in towards the M-line, bringing the Z-discs closer together.

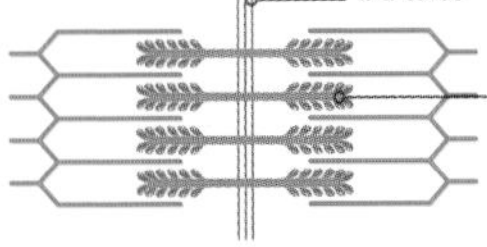

M line

Thick filament

RELAXED SARCOMERE

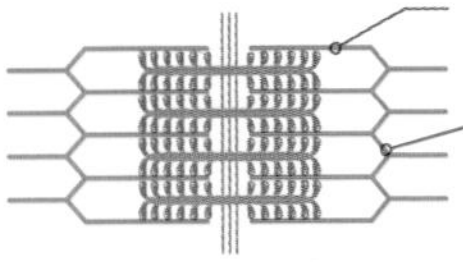

Thin filament

Z disc

CONTRACTED SARCOMERE

NERVOUS
SYSTEM

The nervous system is a control network that connects all body systems. It is split into the central and peripheral nervous system (PNS). The PNS is comprised of the somatic and autonomic nervous system.

SYSTEM OVERVIEW

The somatic nervous system consists of nerves carrying sensory and motor signals to and from the spinal cord and brain. The autonomic nervous system (ANS) is divided into two functional systems: the sympathetic nervous system and parasympathetic nervous system, which accounts for many of yoga's benefits.

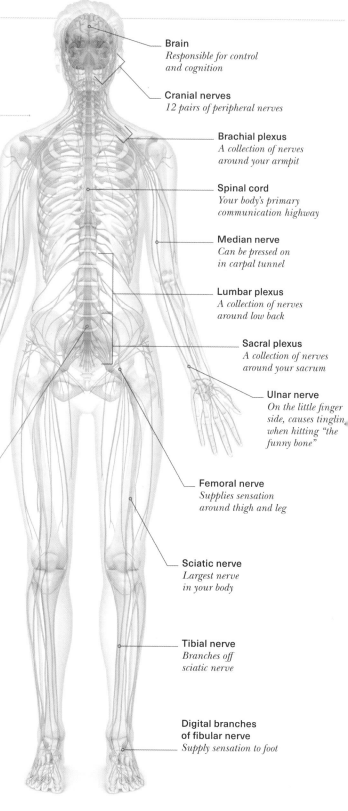

Brain
*Responsible for control
and cognition*

Cranial nerves
12 pairs of peripheral nerves

Brachial plexus
*A collection of nerves
around your armpit*

Spinal cord
*Your body's primary
communication highway*

Median nerve
*Can be pressed on
in carpal tunnel*

Lumbar plexus
*A collection of nerves
around low back*

Sacral plexus
*A collection of nerves
around your sacrum*

Ulnar nerve
*On the little finger
side, causes tinglin
when hitting "the
funny bone"*

Femoral nerve
*Supplies sensation
around thigh and leg*

Sciatic nerve
*Largest nerve
in your body*

Tibial nerve
*Branches off
sciatic nerve*

**Digital branches
of fibular nerve**
Supply sensation to foot

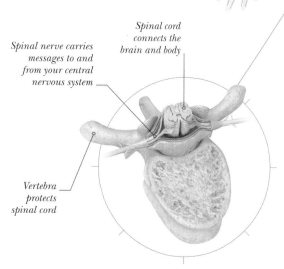

*Spinal cord
connects the
brain and body*

*Spinal nerve carries
messages to and
from your central
nervous system*

*Vertebra
protects
spinal cord*

Spinal cord
In this superior, or bird's eye, view of
a vertebra, you can see how your spinal cord
is protected by the bony encasement of the
spinal column. Spinal nerves project off to
the side, in between the vertebrae.

NERVE STRUCTURE

Neurons are the main cells of your nervous system. Axons are bundled together in your PNS to make nerves. Nerves are like highly conductive electrical wires sending signals throughout your body. Some are wrapped with a fatty substance called myelin, making their signals travel faster.

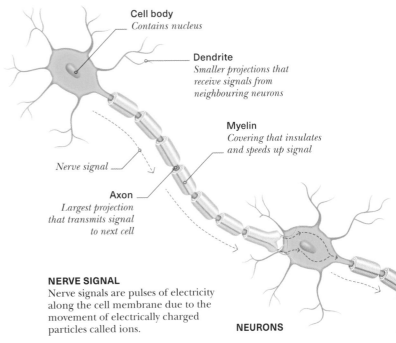

Cell body
Contains nucleus

Dendrite
Smaller projections that receive signals from neighbouring neurons

Myelin
Covering that insulates and speeds up signal

Nerve signal

Axon
Largest projection that transmits signal to next cell

NEURONS

NERVE SIGNAL
Nerve signals are pulses of electricity along the cell membrane due to the movement of electrically charged particles called ions.

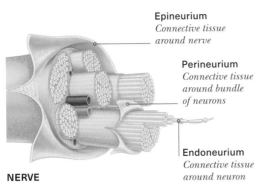

Epineurium
Connective tissue around nerve

Perineurium
Connective tissue around bundle of neurons

Endoneurium
Connective tissue around neuron

NERVE

THE AUTONOMIC NERVOUS SYSTEM

The autonomic nervous system (ANS) can be thought of as your body's autopilot. Its functions are automatic and they include processes such as your heart rate, breathing, digestion, and excretion, which happen without you having to consciously think about them. The ANS is further divided into two systems of control that complement each other: the sympathetic nervous system (SNS) and the parasympathetic nevous system (PSNS).

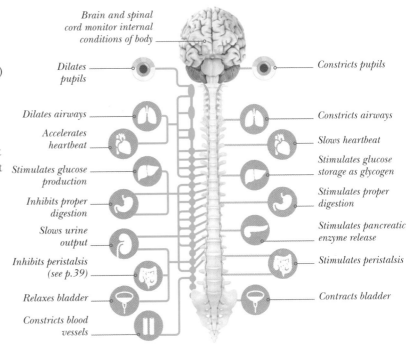

Brain and spinal cord monitor internal conditions of body

Dilates pupils

Constricts pupils

Dilates airways

Constricts airways

Accelerates heartbeat

Slows heartbeat

Stimulates glucose production

Stimulates glucose storage as glycogen

Inhibits proper digestion

Stimulates proper digestion

Slows urine output

Stimulates pancreatic enzyme release

Inhibits peristalsis (see p.39)

Stimulates peristalsis

Relaxes bladder

Contracts bladder

Constricts blood vessels

SYMPATHETIC NERVOUS SYSTEM
The SNS is known as "fight or flight" or the "stress response" because it helps you deal with stressful situations.

PARASYMPATHETIC NERVOUS SYSTEM
The PSNS is known as "rest and digest" or the "relaxation response" because it creates a restful state of optimal function.

CEREBRAL CORTEX

Compared to other mammals, our brains are massive for our bodies, with a particularly developed cerebral cortex. Most of the cortex is on the outside of the brain, except the insula. It is composed of grey matter, which is filled with synapses or connection points between neurons. Your cortex has five lobes and many functional areas.

LOBES OF THE BRAIN
The brain is separated into five main divisions, called lobes, including the insula which is inside the brain (not seen here).

INSIDE THE BRAIN

The brain contains many different structures and scientists are still working out what their functions are. Some of these structures monitor conditions inside your body and relay information. The limbic system is the emotional centre of your brain.

INTERNAL STRUCTURES
This image shows the brain as if it were cut in half down the middle (a mid-sagittal section) to reveal structures inside the cerebrum.

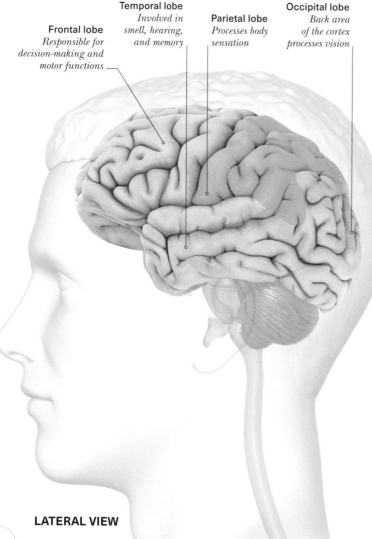

Temporal lobe
Involved in smell, hearing, and memory

Frontal lobe
Responsible for decision-making and motor functions

Parietal lobe
Processes body sensation

Occipital lobe
Back area of the cortex processes vision

LATERAL VIEW

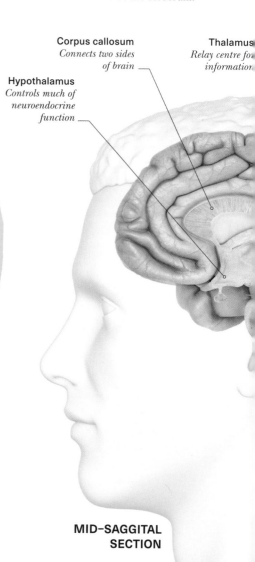

Corpus callosum
Connects two sides of brain

Hypothalamus
Controls much of neuroendocrine function

Thalamus
Relay centre for information

MID-SAGGITAL SECTION

How yoga affects your brain

This chart looks at the neuroscience that may explain the vast mental and physical benefits of yoga. Modern science shows us that the brain maintains its ability to adapt across a lifetime, making it possible to break bad habits and negative patterns. It can also create the key chemicals that pharmaceutical companies synthesize in a lab. Research is uncovering the huge potential of yoga therapy to help people on a global scale. These effects stem from yoga's multidimensional approach, reflected in its 8-limb structure (see p.198), which includes guidelines on self-control and self-regulation.

↑ **Brain alpha wave activity increased** Alpha waves are associated with relaxation

↑ **GABA increased** Gamma-aminobutyric acid counteracts anxiety and stress symptoms, leading to more relaxation.

↑ **Serotonin increased** Serotonin helps regulate your mood. Low levels of usable serotonin are associated with depression.

↑ **BDNF increased** Brain-derived neurotrophic factor is a protein responsible for neuron health and neuroplasticity. Yoga can boost levels of BDNF, which may help people with chronic pain or depression.

⟳ **Dopamine regulated** Dopamine acts as your body's reward system and dysfunction is associated with addiction. Research suggests that meditation results in improved self-regulation.

↓ **Cortisol reduced** Cortisol is a stress hormone. When your baseline increases and levels are too high for too long, it can lead to inflammation and weight gain.

↓ **Norepinephrine reduced** A decrease in norepinephrine, or adrenaline, means fewer stress hormones in your system.

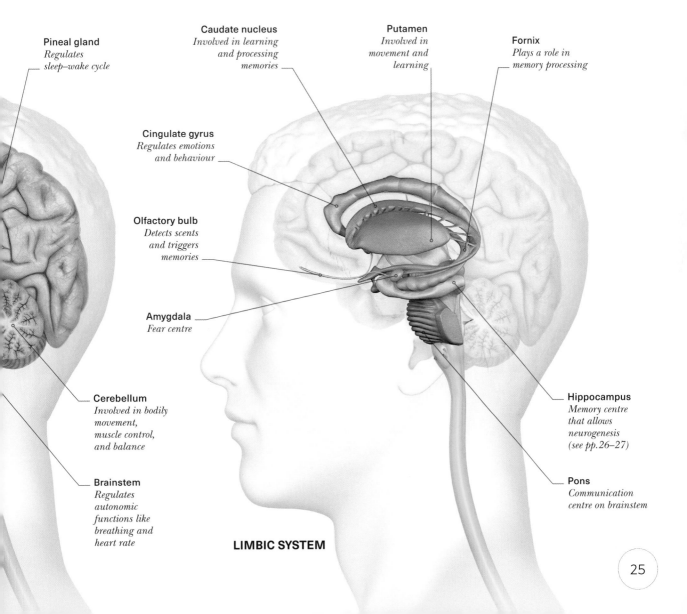

Pineal gland
Regulates sleep–wake cycle

Caudate nucleus
Involved in learning and processing memories

Putamen
Involved in movement and learning

Fornix
Plays a role in memory processing

Cingulate gyrus
Regulates emotions and behaviour

Olfactory bulb
Detects scents and triggers memories

Amygdala
Fear centre

Cerebellum
Involved in bodily movement, muscle control, and balance

Brainstem
Regulates autonomic functions like breathing and heart rate

Hippocampus
Memory centre that allows neurogenesis (see pp.26–27)

Pons
Communication centre on brainstem

LIMBIC SYSTEM

NEURAL PATHWAYS

The brain develops neural connections – and eventually becomes conditioned – based on your choices and experiences. It is said that neurons that fire together, wire together. The more you practise an activity – or a mindset – the more networks are created. With approximately 100 billion neurons, the brain's possible connections are vast. Yoga practices facilitate this process.

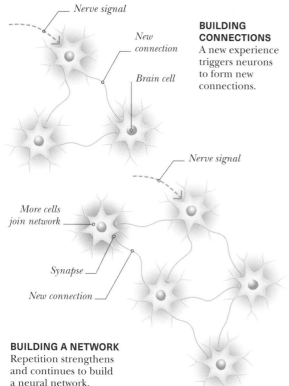

Nerve signal

New connection

Brain cell

BUILDING CONNECTIONS
A new experience triggers neurons to form new connections.

Nerve signal

More cells join network

Synapse

New connection

BUILDING A NETWORK
Repetition strengthens and continues to build a neural network.

CHANGING BRAIN

Neuroplasticity is the ability of your brain to be moulded. Not long ago, scientists thought the brain couldn't change after childhood and degraded with age. Now we know that nervous tissue adapts. Like exercise affects your muscles, your brain tissue either develops or atrophies based on stimulation.

UNSTIMULATED BRAIN
Without stimulation, fewer connections are made. The brain tissue looks like a dying tree with sparse branches.

STIMULATED BRAIN
With stimulation, more connections form. The brain tissue looks like a thriving tree with dense branches.

Samskara

Yogis perhaps conceptualized neuroplasticity with *samskaras*: impressions due to past thoughts and actions. Yoga can help beat bad habits or conditioned responses by affecting neural pathways and *samskaras*. This occurs at a synaptic level each time you consciously change your thoughts and actions through awareness and practice. The more you travel that new path, the stronger the connection between the neurons gets.

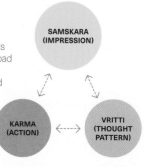

SAMSKARA (IMPRESSION)

KARMA (ACTION)

VRITTI (THOUGHT PATTERN)

CYCLIC NATURE OF HABITS

How yoga boosts your brain

There is no neuroplasticity pill. The most effective way to shape your brain is through behavioural changes. Although any yoga practice should encourage neuroplasticity, try the tips here for improved results.

Up the intensity
Moderate to vigorous physical activity, like from Sun salutations, is one of the most effective ways of increasing brain-derived neurotrophic factor. This is a nerve growth factor, which is like a glue that helps to wire in neural connections.

Change your routine
Purposefully and consciously changing your yoga practice routine benefits your mind and your body.

Meditate
Research shows that meditation builds grey matter in your cerebral cortex.

Join a class
The act of moving with a group and following the teacher activates mirror neurons. The mirror neuron system is a recently discovered network of nerves involved in emulation of movement and developing compassion.

NEUROGENESIS

Scientists used to think that people are born with a certain number of nerve cells and that they cannot grow new ones. Research has since revealed that the growth of new neurons, or neurogenesis, can happen at any age. Neurogenesis occurs in key areas of the brain responsible for memory – the hippocampus – and smell. Neural stem cells in these regions of the brain develop new neurons.

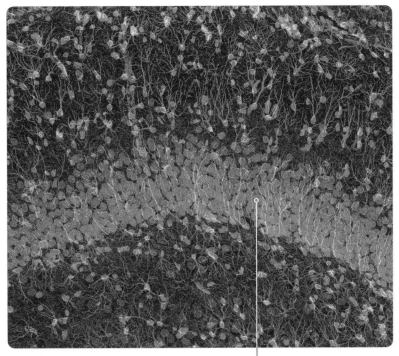

SITE FOR NEW CELLS
In this hippocampus tissue, helper cells or neuroglia are blue, axons are green, and neuron cell bodies and stem cells are pink.

Stem cells
Hippocampal stem cells can develop into new neurons, improving memory

CORTISOL
LEVELS

Consistently high levels of the stress hormone cortisol are related to increased amygdala (fear centre, see p.25) activity and decreased hippocampal (memory centre) activity. When under these conditions, the hippocampus doesn't grow new neurons or connections well. Yoga practices are shown to reduce cortisol levels and reverse these effects, which may contribute to improving memory.

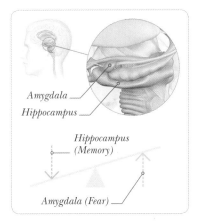

Amygdala
Hippocampus

Hippocampus (Memory)

Amygdala (Fear)

STRESS AND MEMORY
Increased activity in the amygdala is correlated with reduced activity in the hippocampus, which has an adverse effect on memory.

Practise hand mudras
Hand mudras are gestures that require concentration and awareness. Just as people who read Braille have more developed hand-specific sensory areas of their brain, mudras may develop brain areas linked with sensory acuity, and fine motor skills.

PADMA MUDRA

HAKINI MUDRA

SHUNI MUDRA

BUDDHI MUDRA

ENDOCRINE
SYSTEM

The endocrine system is a slower, longer lasting control system than the nervous system. It consists of glands that release hormones into your bloodstream to be delivered to specific cells.

SYSTEM OVERVIEW

Your brain controls the release of hormones from endocrine glands to maintain a balance inside your body, called homeostasis. Stressors – from external environmental conditions to internal or emotional factors – affect this balance but yoga can help. For example, research suggests that yoga may prevent and improve symptoms of type 2 diabetes.

Pineal gland
Makes melatonin, which affects sleep

Hypothalamus
Controls other glands

Pituitary gland
Produces key hormones

Parathyroid gland
Regulates blood calcium levels

Thyroid gland
Regulates metabolism and blood calcium

Heart
Releases hormones to regulate blood pressure

Suprarenal gland
Regulates salt levels and produces adrenaline in response to danger

Pancreas
Secretes insulin and glucagon to regulate blood sugar

Small intestine
Releases hormones to help with digestion

Ovaries produce female sex hormones

Testis
Produces male sex hormones

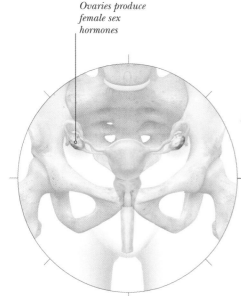

FEMALE

MALE

Homeostasis and allostasis

Homeostasis is your body's state of dynamic equilibrium. Most processes – like the control of hormone release, blood calcium and blood sugar levels, and temperature – are tightly regulated through negative feedback, which works in a similar way to a thermostat. Nature wants you to be in balance. Yogis referred to this as *samatva* which can be translated as equilibrium or equanimity. Allostasis is a process of maintaining homeostasis amidst stressors. The more intense the stress, the heavier the "allostatic load" and the more your cells have to work to maintain equilibrium. This increases the likelihood of chronic diseases. Researchers believe that yoga can reduce allostatic load.

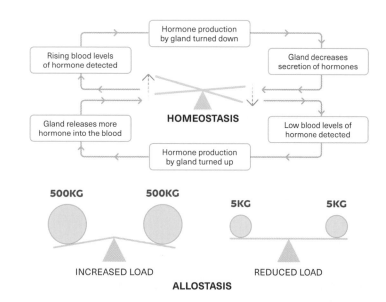

Hormone production by gland turned down

Rising blood levels of hormone detected

Gland decreases secretion of hormones

HOMEOSTASIS

Gland releases more hormone into the blood

Low blood levels of hormone detected

Hormone production by gland turned up

500KG   500KG   5KG   5KG

INCREASED LOAD   REDUCED LOAD

ALLOSTASIS

PANCREAS

Your pancreas releases insulin to help sugar get into your body cells. However, cells can become insulin-resistant causing disease. A review found that yoga can improve glycaemic control, lipid levels, and body composition of fat in those with metabolic syndrome and type 2 diabetes. A doctor-approved reduction in medications was also found.

Metabolism

Most yoga practices tend to slow your metabolism, helping your body to be more efficient with less. Although your metabolism may slightly lower from relaxation-based practices, this doesn't mean you will gain weight. A reduction in stress hormones like cortisol also prevents your body from holding onto fat.

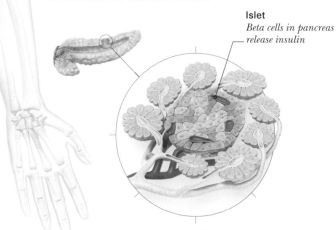

Islet
Beta cells in pancreas release insulin

Pancreatic islets
Inside your pancreas, islets contain different types of cells. Beta cells release insulin, which allows your body cells to use glucose.

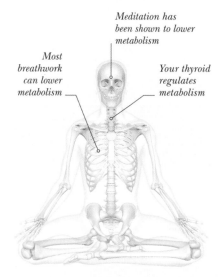

Meditation has been shown to lower metabolism

Most breathwork can lower metabolism

Your thyroid regulates metabolism

RESPIRATORY
SYSTEM

You take a breath 12–20 times per minute. The purpose of your breath is to get oxygen to your cells and to get rid of waste like carbon dioxide. The respiratory system includes the nasal cavities, air passageway tubes, and lungs.

SYSTEM OVERVIEW

You don't have to think to breathe; respiration is a part of your autonomic nervous function. However, yogis claim that by controlling your breath, you can control all aspects of being. Science reveals that your breath is an access point to regulating your nervous system.

 Neti pots

The neti pot is a part of traditional yogic hygiene practices. It involves pouring clean (filtered or boiled), warm salt water in one nostril to fill the sinuses and drain out the other nostril. Neti pots (or a similar sinus rinse) are recommended by many modern physicians to help allergies and respiratory illness.

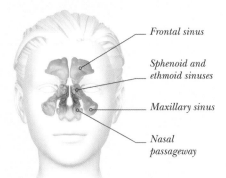

Frontal sinus

Sphenoid and ethmoid sinuses

Maxillary sinus

Nasal passageway

SINUSES
Your sinuses are a system of connected, air-filled cavities in your skull. They make your skull lighter, help your voice to resonate, and affect your breath.

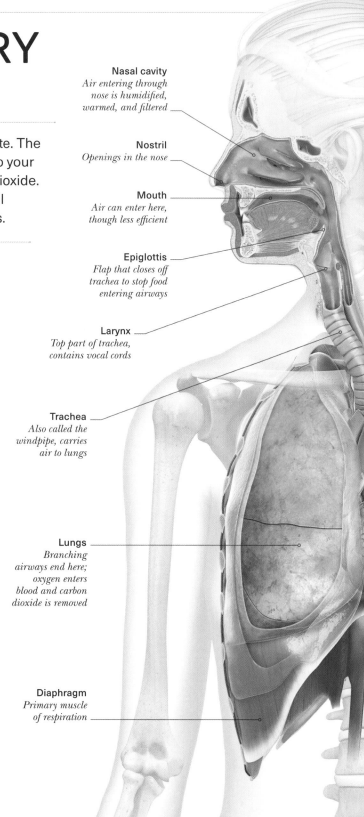

Nasal cavity
Air entering through nose is humidified, warmed, and filtered

Nostril
Openings in the nose

Mouth
Air can enter here, though less efficient

Epiglottis
Flap that closes off trachea to stop food entering airways

Larynx
Top part of trachea, contains vocal cords

Trachea
Also called the windpipe, carries air to lungs

Lungs
Branching airways end here; oxygen enters blood and carbon dioxide is removed

Diaphragm
Primary muscle of respiration

HOW WE **BREATHE**

When you inhale, the breath enters your nose, throat, and then your lungs. Your lungs and ribcage expand three-dimensionally in all directions; your diaphragm engages to flatten down. When you exhale, your diaphragm relaxes to ascend, your lungs and ribcage compress, and the air releases out of your throat and then nose.

Ribcage
Bones surrounding lungs

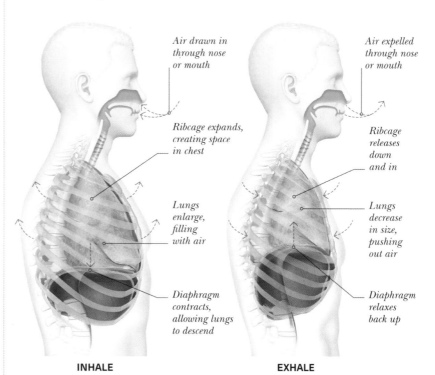

Air drawn in through nose or mouth

Ribcage expands, creating space in chest

Lungs enlarge, filling with air

Diaphragm contracts, allowing lungs to descend

INHALE

Air expelled through nose or mouth

Ribcage releases down and in

Lungs decrease in size, pushing out air

Diaphragm relaxes back up

EXHALE

Belly breathing

"Belly breathing" doesn't mean you are actually breathing in your belly, but rather that you are allowing your belly to move freely with your breath. When your diaphragm engages with the inhale, it presses against your abdominal organs – pushing down and out, which is why this is also called diaphragmatic breathing.

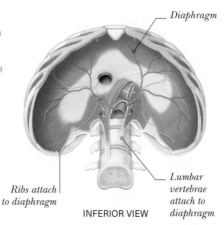

Diaphragm

Ribs attach to diaphragm

Lumbar vertebrae attach to diaphragm

INFERIOR VIEW

PRANAYAMA

Yogis use *pranayama* or breathwork to control their *prana* and anchor to the present moment. The word *prana* in Sanskrit means vital energy or life-force energy that permeates through us and everything. Interestingly, *prana* simultaneously means breath. Yogis believe that you can change the flow and qualities of your energetics by breath control.

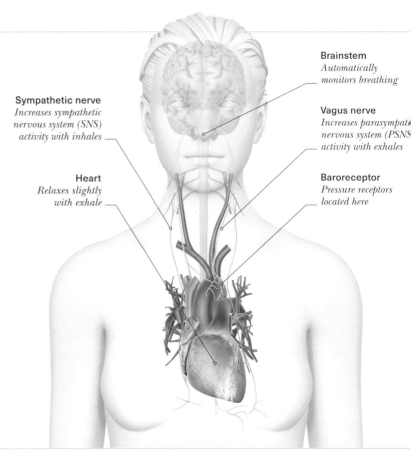

Sympathetic nerve
Increases sympathetic nervous system (SNS) activity with inhales

Heart
Relaxes slightly with exhale

Brainstem
Automatically monitors breathing

Vagus nerve
Increases parasympathetic nervous system (PSNS) activity with exhales

Baroreceptor
Pressure receptors located here

INHALE AND EXHALE

When you inhale, blood is shunted to your heart and lungs to help them function. Baroreceptors (see p.134) sense this increased pressure and respond by signalling to let off the brake pedal, increasing sympathetic activity momentarily. During each exhale, your heart is slightly more relaxed with increased parasympathetic activity. This explains why elongating your exhales in *pranayama* is relaxing.

BREATHWORK
PRACTICES

Modern yogis use breathwork for health benefits, including overcoming inefficient breathing patterns from a culture of poor posture and stress. Through altering your breath, you change your state of mind. For example, you may practise left nostril breathing and bee breath to calm down or right nostril breathing and *kapalabhati* for alertness.

BREATH OF FIRE (KAPALABHATI)

This is a fast breath that mimics hyperventilation, increasing your heart rate and blood pressure. It also tones your abdominals. Avoid this technique if you are pregnant or have anxiety, certain eye conditions, or high blood pressure. Similar effects and precautions apply for breath holding (*kumbhaka*).

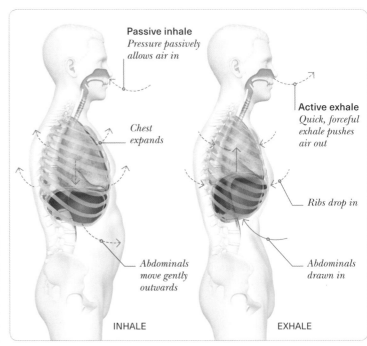

Passive inhale
Pressure passively allows air in

Chest expands

Abdominals move gently outwards

Active exhale
Quick, forceful exhale pushes air out

Ribs drop in

Abdominals drawn in

INHALE

EXHALE

NASAL CYCLE

For many, each nostril takes turns dominating air flow (in .5- to 4-hour shifts). This is called the nasal cycle. You probably notice this more when you are congested. Openness indicates local vasoconstriction and the swollenness indicates vasodilation. Observe this cycle naturally or try purposefully covering one nostril for a desired effect (see panel right).

Left brain, right brain

Each half of your body is controlled by the opposite hemisphere of your brain – meaning that your left arm is controlled by the right half of your brain. The same is true of your nostrils. This may have many implications, including a slight overall increase in SNS activity when right nostril breathing and PSNS when left, although evidence is mixed.

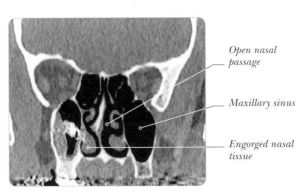

Open nasal passage

Maxillary sinus

Engorged nasal tissue

NASAL TISSUE
This image shows the right nasal passage swollen while the left is open. In this case the swelling is exacerbated by congestion.

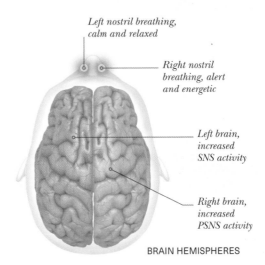

Left nostril breathing, calm and relaxed

Right nostril breathing, alert and energetic

Left brain, increased SNS activity

Right brain, increased PSNS activity

BRAIN HEMISPHERES

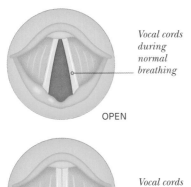

Vocal cords during normal breathing

OPEN

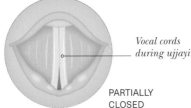

Vocal cords during ujjayi

PARTIALLY CLOSED

VICTORIOUS BREATH (UJJAYI)
Victorious breath involves constricting your vocal cords partially. This feels like when you whisper softly. It creates an ocean sound to give your mind a focal point.

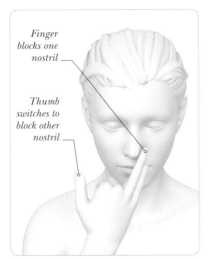

Finger blocks one nostril

Thumb switches to block other nostril

ALTERNATE NOSTRIL BREATHING
This technique may calm the mind and body. It involves focus and activation of both sides of the brain. To practise it just remember: exhale, inhale, and switch nostrils.

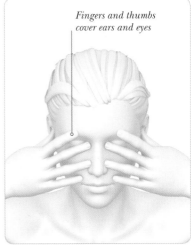

Fingers and thumbs cover ears and eyes

BEE BREATH (BRAHMARI)
This involves covering your eyes and ears and humming on a long exhale. Yogis used this to improve sleep. Research shows it can lower heart rate, blood pressure, and anxiety.

CARDIOVASCULAR
SYSTEM

The heart, an intricate network of vessels, and the blood circulating through them make up your cardiovascular system.

SYSTEM OVERVIEW

Your heart constantly beats to pump blood around your body, removing waste and delivering vital oxygen. Research on yoga suggests profound benefits for cardiovascular health, including reduced risk of heart disease. Yoga has been shown to clinically improve blood pressure, cholesterol levels, and cardiovascular resilience (see opposite).

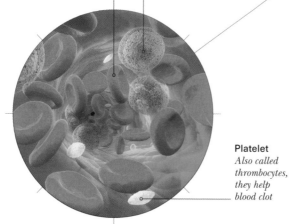

Red blood cell
Also called erythrocytes, they deliver oxygen

White blood cell
Also called leukocytes, they fight invaders

Platelet
Also called thrombocytes, they help blood clot

Composition of blood
Adults have about 5 litres (11 pt) of blood circulating around the body. Blood is a connective tissue composed of red blood cells, white blood cells, and platelets suspended in a liquid called plasma. It provides oxygen, nutrients, and hormones as well as removing waste from cells.

Jugular vein
Returns blood from brain to heart

Carot.. arter.
Deliv.. blood.. brain

Subclavian artery
Carries blood to arm and hand

Superior vena cava
Returns blood back to heart

Aorta
Largest artery in your body

Heart
Muscular pump for blood

Inferior vena cava
Returns blood from lower body to heart

Abdominal aorta
Delivers blood to abdomen and lower body

Femoral vein
Delivers blood from lower limb to heart

Femoral artery
Carries blood to thighs

Popliteal artery
Carries blood to knee and calf

Great saphenous vein
Longest vein in the body

HEART AND CIRCULATION

Circulation has two loops: pulmonary (lungs) and systemic (body). Veins carry blood to the heart, and arteries carry it away. Veins are shown in blue to represent deoxygenation, and arteries are red for oxygenation. The exceptions are pulmonary arteries (deoxygenated) and pulmonary veins (oxygenated).

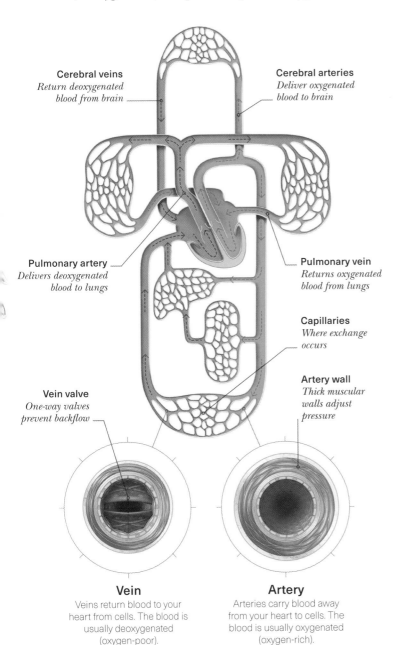

Cerebral veins
Return deoxygenated blood from brain

Cerebral arteries
Deliver oxygenated blood to brain

Pulmonary artery
Delivers deoxygenated blood to lungs

Pulmonary vein
Returns oxygenated blood from lungs

Capillaries
Where exchange occurs

Artery wall
Thick muscular walls adjust pressure

Vein valve
One-way valves prevent backflow

Vein
Veins return blood to your heart from cells. The blood is usually deoxygenated (oxygen-poor).

Artery
Arteries carry blood away from your heart to cells. The blood is usually oxygenated (oxygen-rich).

Heart rate variability

Heart rate variability (HRV) is the heart's ability to adapt fast. It is better for your pulse to vary rather than tick steadily. High HRV shows autonomic resilience and may lead to improved physical, emotional, and cognitive function. Yoga appears to improve HRV.

HEARTBEAT

Hypertension

Research shows that yoga can reduce blood pressure significantly. With over 1 billion people living with hypertension, yoga offers a cost-effective adjunct to care with minimal to no side-effects. Consult your doctor about any blood pressure shifts.

BLOOD PRESSURE MONITOR

Cholesterol

Reports have shown that yoga can increase "good" cholesterol (high-density lipoprotein or HDL) and decrease "bad" cholesterol (low-density lipoprotein or LDL). This reduces the risk of heart disease by preventing arterial narrowing.

NARROWED ARTERY

Heart disease

A meta-analysis suggests that yoga reduces heart disease risk as well as or better than accepted exercise guidelines. Long-term clinical trials have shown that a yogic lifestyle – with asanas, meditation, social support, and a plant-based diet – could reverse heart disease.

DAMAGED HEART TISSUE

LYMPHATIC
SYSTEM

The lymphatic and immune systems work together to fight invaders. Acute inflammation can be a helpful result of this internal war, such as when you have a cut. However, chronic inflammation is an underlying cause of many major diseases.

SYSTEM OVERVIEW

Lymph vessels collect and drain excess fluid from body tissues. They also carry immune cells around your body. Evidence suggests that yoga can help reduce chronic inflammation and it may boost immunity, helping you get sick less often and less intensely. Your body can heal itself and yoga can help.

Tonsils
Help destroy bacteria or viruses that enter the nose or mouth

Thoracic duct
Lymph drains back into the heart through here

Axillary nodes
A concentration of lymph nodes under the arm

Spleen
Produces cells that fight infection

Cisterna chyli
Collects lymph from the lower half of the body

Inguinal nodes
A concentration of lymph nodes around the groin

Lymph node
Lymph is processed and cleaned here

Lymph vessel
Drains and transports lymph

Valve keeps lymph flowing in one direction

Lymph flows out of node

Lymphocytes, specialized white blood cells

Lymph node
These are checkpoints that screen lymphatic fluid for foreign invaders. The cleaned fluid is returned to your blood. Movement in yoga asanas, particularly from sun salutations and inversions, can help facilitate lymph flow.

White blood cells

White blood cells are like warriors fighting viruses, bacteria, and cancer cells in your body. Fragments of the invaders, called antigens, are presented so the warriors can strategically fight using the right antibodies and chemical messengers called cytokines. Communication is key – miscommunication can lead to chronic inflammation.

DENDRITIC CELL
These present antigens, which the body recognizes as a foreign invader. They activate T-cells to do their job.

MACROPHAGE
Hungry hunter cells (see phagocytosis below) that also release cytokines to induce inflammation.

B-CELL
A type of lymphocyte that secretes antibodies, which are proteins specialized to fight specific antigens.

T-CELL
A type of lymphocyte that is activated to fight by the presentation of antigens. There are many specialized types.

PHAGOCYTOSIS
Macrophages (white) patrol your body on alert for invaders (red) to engulf and eat in a process called phagocytosis.

INFLAMMATORY RESPONSE

Inflammation often involves heat, pain, redness, and swelling due to a cascade of events where white blood cells fight invaders. In an autoimmune disease, they mistakenly fight body tissue. For example, rheumatoid arthritis (see below) can flare to cause local inflammation and body-wide inflammation.

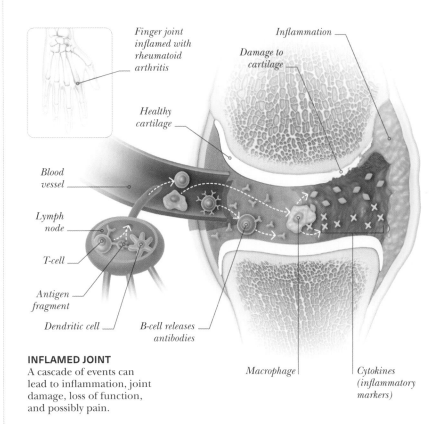

Finger joint inflamed with rheumatoid arthritis

Inflammation

Damage to cartilage

Healthy cartilage

Blood vessel

Lymph node

T-cell

Antigen fragment

Dendritic cell

B-cell releases antibodies

Macrophage

Cytokines (inflammatory markers)

INFLAMED JOINT
A cascade of events can lead to inflammation, joint damage, loss of function, and possibly pain.

Yoga and inflammation
Yoga seems to help attenuate inflammation by reducing the stress response, which may reduce your disease risk. A review shows that yoga practice reduces cytokine count and therefore inflammation. Scientists hypothesize that a long-term, regular practice would be most effective.

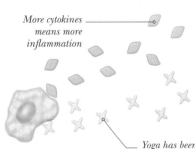

More cytokines means more inflammation

Yoga has been shown to reduce cytokines, including: IL-1beta, TNF-alpha, IL6, and IL10

CYTOKINES
These are inflammatory markers that encourage an immune response.

DIGESTIVE
SYSTEM

The digestive tract is a tube with selective membranes that control what gets into your body. Nutrients are absorbed and waste is expelled.

SYSTEM OVERVIEW

Food is broken down into absorbable units by your digestive system, from chewing in the mouth to chemical breakdown in the stomach and squeezing in the intestines. Nutrients enter the blood, and ultimately your cells. Yogis recognized that you become what you eat, equating the physical body (*anamaya*) with the "food body".

Journey of food

It is best to practise yoga asanas with an empty stomach. That may mean not eating a meal 2–4 hours before class. You may need to strategically plan a small snack, especially if you tend to have low blood sugar or other medical conditions.

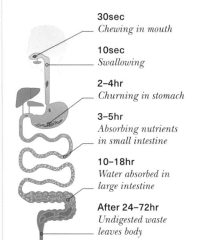

30sec
Chewing in mouth

10sec
Swallowing

2–4hr
Churning in stomach

3–5hr
Absorbing nutrients in small intestine

10–18hr
Water absorbed in large intestine

After 24–72hr
Undigested waste leaves body

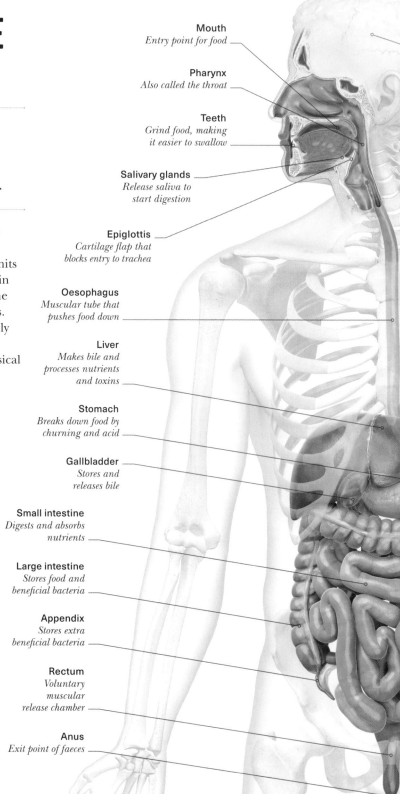

Mouth
Entry point for food

Pharynx
Also called the throat

Teeth
Grind food, making it easier to swallow

Salivary glands
Release saliva to start digestion

Epiglottis
Cartilage flap that blocks entry to trachea

Oesophagus
Muscular tube that pushes food down

Liver
Makes bile and processes nutrients and toxins

Stomach
Breaks down food by churning and acid

Gallbladder
Stores and releases bile

Small intestine
Digests and absorbs nutrients

Large intestine
Stores food and beneficial bacteria

Appendix
Stores extra beneficial bacteria

Rectum
Voluntary muscular release chamber

Anus
Exit point of faeces

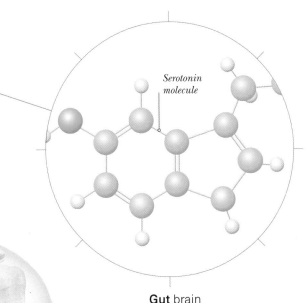

Serotonin
molecule

Enteric nervous system (ENS)

Scientists have recently discovered the semi-independent enteric nervous system (ENS). These 100 million neurons may be responsible for you feeling butterflies in your stomach from love or an intuitive gut feeling. Yoga enhances your mind-body connection, so you can feel what is going on in your gut clearly. This interconnection may explain how yoga can improve both your digestion and mood significantly.

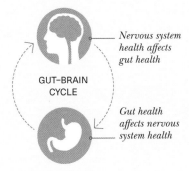

Nervous system health affects gut health

GUT–BRAIN CYCLE

Gut health affects nervous system health

Gut brain

Around 95% of your serotonin, a chemical needed for mood regulation, is stored in and partially controlled by your gut. "Gut brain" or enteric nervous system (see right) dysfunction is associated with gastrointestinal upset and irritable bowel syndrome (IBS), depression, and anxiety.

Ahimsa diet

Yogis often make conscious choices about what they put into their body. An ahimsa diet is one of non-harm. For many, this means being a vegetarian to reduce the suffering of other animals. A largely plant-based diet reduces your risk of heart disease, cancer, and related major killers. Scientists project that a mostly vegetarian diet may reduce global mortality by 6–10% and cut food-based greenhouse gas emissions by 29–70% – a huge impact on the environment. Even small dietary changes like a Meatless Monday can make a big difference.

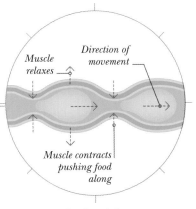

Muscle relaxes

Direction of movement

Muscle contracts pushing food along

Peristalsis

Peristalsis is the involuntary smooth muscular movement of foodstuff through your digestive tract. It's encouraged by the relaxation response and physical movement, as from yoga asana practice.

NON-HARM FOOD

URINARY
SYSTEM

The urinary system filters out waste and excess fluids to maintain correct blood volume. This in turn affects blood pressure, which yoga has also been shown to help regulate.

SYSTEM OVERVIEW

Your kidneys process waste from blood into urine, which is then stored in your bladder. Urine release is voluntary in adults but some people lose this control, leading to urinary incontinence. A recent study showed that yoga classes may help manage urinary incontinence.

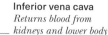

Inferior vena cava
Returns blood from kidneys and lower body

Abdominal aorta
Delivers blood to kidneys and lower body

Adrenal gland
Regulates fluid volume

Kidney
Filters blood to make urine

Ureter
Carries urine from kidneys to bladder

Bladder
Stores urine

Prostate gland
Surrounds male urethra

Urethra
Carries urine from bladder out of body

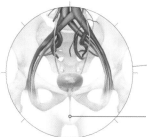

Urethra
Shorter urethra increases chance of infections

FEMALE

MALE

Pelvic floor muscles

Your pelvic floor muscles are vital for bladder control. Common issues such as frequent, urgent, or painful urination, or slight leaking – such as when sneezing or laughing – may be helped by yoga exercises. For example, a gentle version of *mula bandha* (see p.153) and relaxation practices could improve pelvic floor health.

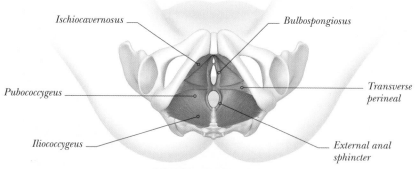

Ischiocavernosus

Bulbospongiosus

Pubococcygeus

Transverse perineal

Iliococcygeus

External anal sphincter

INFERIOR VIEW OF FEMALE

REPRODUCTIVE
SYSTEM

The reproductive system functions to help continue our species by sexual reproduction. Yoga may help aspects of reproductive health, including pelvic floor health. This may improve sexual satisfaction, and labour and delivery.

SYSTEM OVERVIEW

Yoga seems to indirectly address aspects of pelvic health, both urinary and reproductive, partly by promoting optimal breathing. It is also feasible that because yoga helps manage stress, it can improve fertility and conception; although we need more research to confirm this.

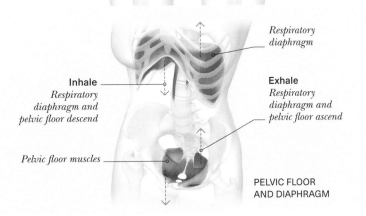

Milk ducts
Carry milk from glands to nipple

Nipple
Opening where a baby sucks milk

Fallopian tube
Connects ovary to uterus

Uterus
Where a fertilized egg develops

Ovary
Where eggs are stored and released

Endometrium
Uterus lining that thickens to receive an egg

Cervix
Opening to uterus

Vagina
Muscular tube

FEMALE

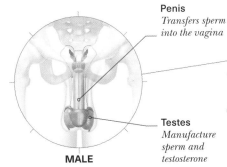

Penis
Transfers sperm into the vagina

Testes
Manufacture sperm and testosterone

MALE

Pelvic floor motion

A healthy pelvic floor is able to move through its full range of motion with your breath, following the movement of your diaphragm. Yoga practice may enhance neurological awareness, along with increasing strength, flexibility, and the relaxation of these muscles. This may improve your bladder, bowel, sexual, and reproductive health.

BREATHING
Your pelvic floor muscles descend as you inhale and ascend as you exhale.

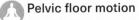

Respiratory diaphragm

Inhale
Respiratory diaphragm and pelvic floor descend

Exhale
Respiratory diaphragm and pelvic floor ascend

Pelvic floor muscles

PELVIC FLOOR
AND DIAPHRAGM

SEATED
Pages 44–83

STANDING
Pages 84–121

INVERSIONS
Pages 122–143

FLOOR
Pages 144–173

THE
ASANAS

Allow this section to guide a meditative exploration of your inner world. Visualize, physically touch, and become curious about how your body feels. Studying these 30 asanas can be an engaging way to memorize the muscles, and better understand the basics of anatomy, physiology, and kinesiology. I hope these poses, or any variation of them, help you become more connected to yourself.

SEATED
ASANAS

Seated and kneeling poses can be grounding and meditative, often forming the starting and ending points of yoga sessions. The asanas presented here show how the body can benefit physically from yoga in a range of ways. Use variations and modify to find stability and ease in body and mind, and remember: if you can breathe, you can do yoga.

ACCOMPLISHED
Siddhasana

This seated pose is so called because the traditional purpose of all the other poses is to prepare your body physically for this meditative posture. The neutral spine and engaged abdominals should make the pose steady and comfortable; if it isn't, try other options.

THE BIG PICTURE

Your back muscles and abdominals engage, while stretching muscles on the outside of your hips. You may feel this minimally, but for many people it can be challenging to maintain a neutral spine and pelvis, involving using muscles in ways your body isn't used to.

KEY

•-- *Joints*

○-- *Muscles*

● Engaging

● Engaging while stretching

● Stretching

Legs crossed comfortably

VARIATION
The common variation *Sukhasana*, or "easy pose", has the legs crossing at the shins. For many, this may not be so "easy"; find support by sitting on a prop to elevate your hips.

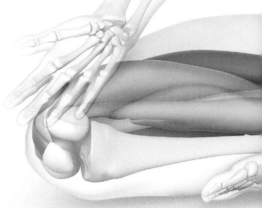

Shoulder
Deltoids

Arms
Your arms relax with the palms of the hands facing up (supinated). Your **posterior deltoid** initiates external shoulder rotation, while your **anterior deltoid** is slightly stretching.

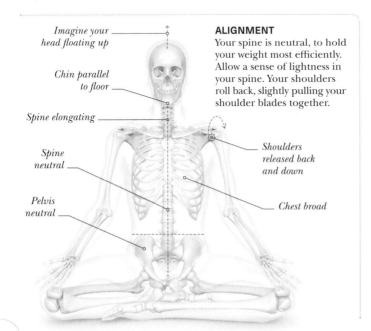

Imagine your head floating up

Chin parallel to floor

Spine elongating

Spine neutral

Pelvis neutral

ALIGNMENT
Your spine is neutral, to hold your weight most efficiently. Allow a sense of lightness in your spine. Your shoulders roll back, slightly pulling your shoulder blades together.

Shoulders released back and down

Chest broad

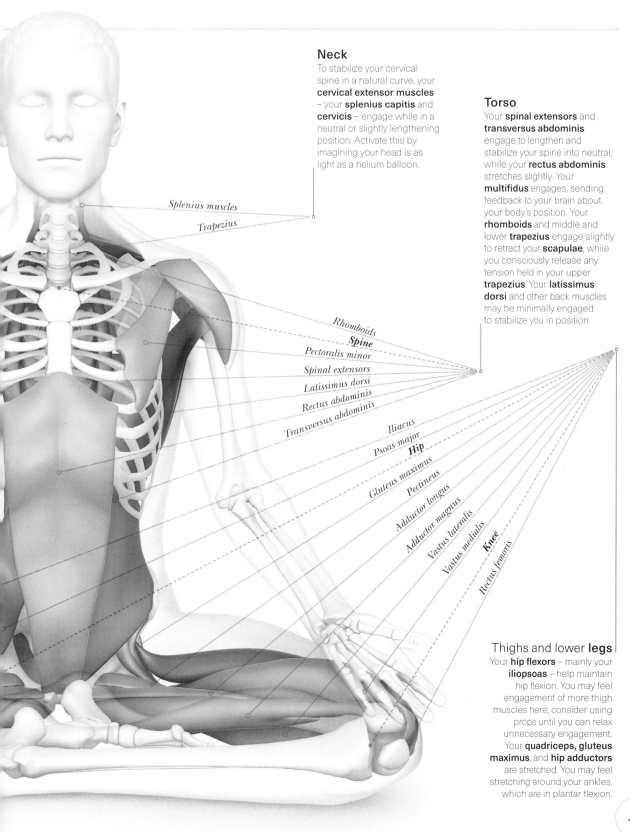

Neck

To stabilize your cervical spine in a natural curve, your **cervical extensor muscles** – your **splenius capitis** and **cervicis** – engage while in a neutral or slightly lengthening position. Activate this by imagining your head is as light as a helium balloon.

Torso

Your **spinal extensors** and **transversus abdominis** engage to lengthen and stabilize your spine into neutral, while your **rectus abdominis** stretches slightly. Your **multifidus** engages, sending feedback to your brain about your body's position. Your **rhomboids** and middle and lower **trapezius** engage slightly to retract your **scapulae**, while you consciously release any tension held in your upper **trapezius**. Your **latissimus dorsi** and other back muscles may be minimally engaged to stabilize you in position.

Splenius muscles

Trapezius

Rhomboids

Spine

Pectoralis minor

Spinal extensors

Latissimus dorsi

Rectus abdominis

Transversus abdominis

Iliacus

Psoas major

Hip

Gluteus maximus

Pectineus

Adductor longus

Adductor magnus

Vastus lateralis

Vastus medialis

Knee

Rectus femoris

Thighs and lower **legs**

Your **hip flexors** – mainly your **iliopsoas** – help maintain hip flexion. You may feel engagement of more thigh muscles here; consider using props until you can relax unnecessary engagement. Your **quadriceps, gluteus maximus**, and **hip adductors** are stretched. You may feel stretching around your ankles, which are in plantar flexion.

» CLOSER LOOK

In Accomplished pose, your intervertebral discs are stacked on top of each other, creating the natural curves of the neutral spine. As you breathe, your ribcage expands and releases efficiently, which is facilitated by sitting tall with good posture.

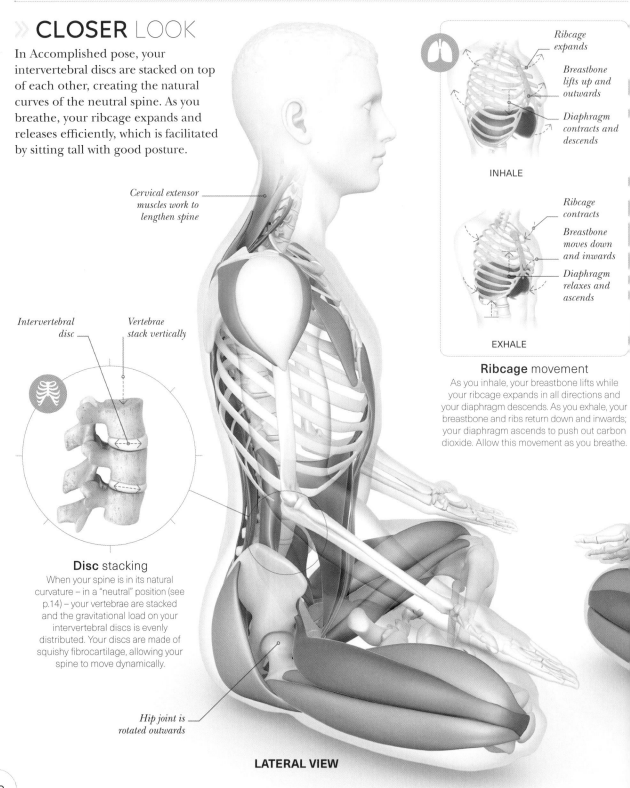

Cervical extensor muscles work to lengthen spine

Intervertebral disc

Vertebrae stack vertically

Disc stacking

When your spine is in its natural curvature – in a "neutral" position (see p.14) – your vertebrae are stacked and the gravitational load on your intervertebral discs is evenly distributed. Your discs are made of squishy fibrocartilage, allowing your spine to move dynamically.

Hip joint is rotated outwards

LATERAL VIEW

Ribcage expands

Breastbone lifts up and outwards

Diaphragm contracts and descends

INHALE

Ribcage contracts

Breastbone moves down and inwards

Diaphragm relaxes and ascends

EXHALE

Ribcage movement

As you inhale, your breastbone lifts while your ribcage expands in all directions and your diaphragm descends. As you exhale, your breastbone and ribs return down and inwards; your diaphragm ascends to push out carbon dioxide. Allow this movement as you breathe.

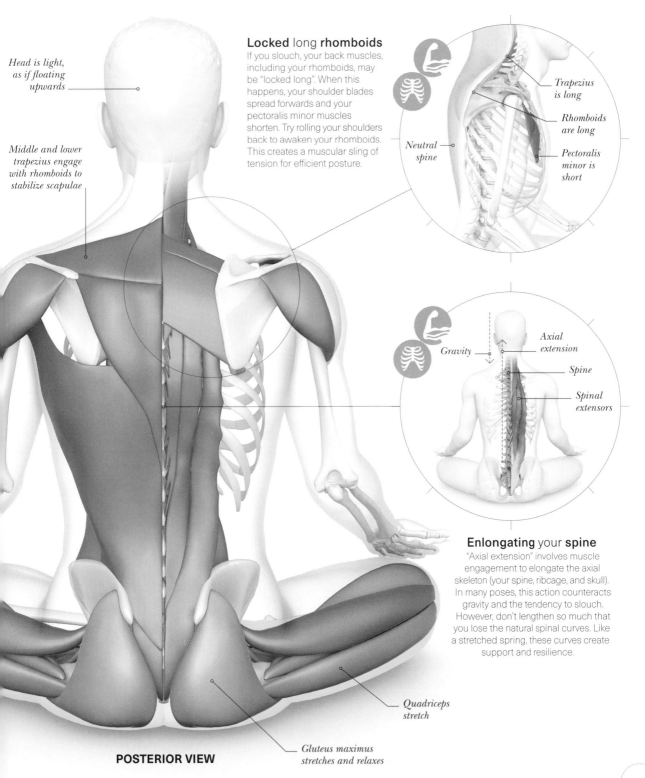

Head is light,
as if floating
upwards

Middle and lower
trapezius engage
with rhomboids to
stabilize scapulae

Locked long rhomboids

If you slouch, your back muscles, including your rhomboids, may be "locked long". When this happens, your shoulder blades spread forwards and your pectoralis minor muscles shorten. Try rolling your shoulders back to awaken your rhomboids. This creates a muscular sling of tension for efficient posture.

Neutral spine

Trapezius is long

Rhomboids are long

Pectoralis minor is short

Gravity

Axial extension

Spine

Spinal extensors

Enlongating your spine

"Axial extension" involves muscle engagement to elongate the axial skeleton (your spine, ribcage, and skull). In many poses, this action counteracts gravity and the tendency to slouch. However, don't lengthen so much that you lose the natural spinal curves. Like a stretched spring, these curves create support and resilience.

Quadriceps stretch

POSTERIOR VIEW

Gluteus maximus stretches and relaxes

49

BOUND ANGLE
Baddha Konasana

Bound Angle pose is a seated hip opener and groin stretch. It can relieve pelvic cramping, and this version of the pose also improves your ankle flexibility and awareness, which will come in handy in balancing poses.

THE BIG PICTURE

Your inner thighs stretch, particularly around your groin. If you can reach, this is also an opportunity to stretch your ankle muscles by opening your feet like a book revealing its pages.

KEY
- *Joints*
- *Muscles*
- Engaging
- Engaging while stretching
- Stretching

Arms
As you reach toward your feet with flexed elbows, your **brachialis** flexes your elbow with the help of the **biceps brachii** and **brachioradialis**.

Shoulder
Biceps brachii
Brachialis
Brachioradialis
Elbow

Lower **legs**
Your **tibialis anterior** muscles dorsiflex your ankles, and your **extensor digitorum** muscles extend your toes. If you are using your hands to manually invert your feet, your **fibularis muscles** are stretching.

Ankle
Fibularis muscles
Extensor digitorum longus
Tibialis anterior

ALIGNMENT
Your spine is stabilized into neutral and, in this version of the pose, your pelvis is also neutral. Your thighs rest in a rotated outward position.

Spine elongated

Shoulders relaxed back and down

Pelvis neutral

Hip rotating outwards

Spine neutral

Feet opened like book

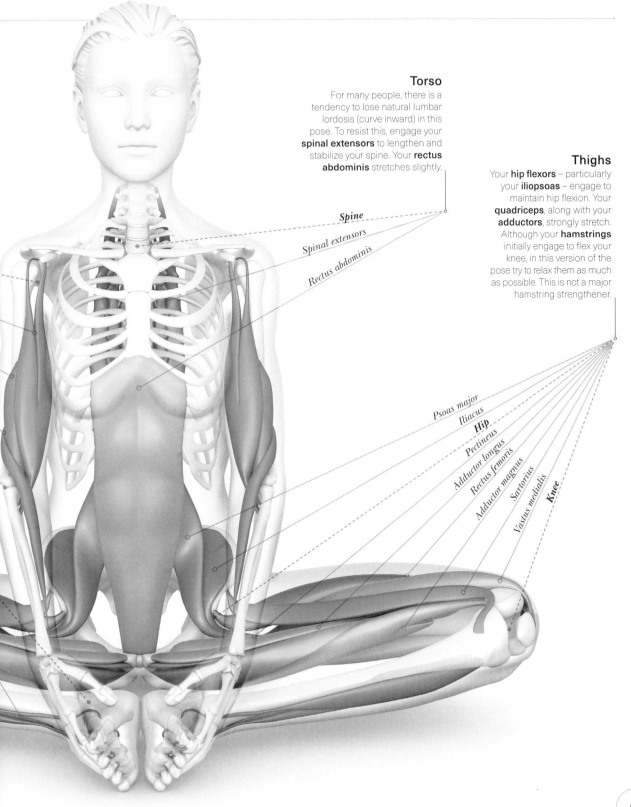

Torso

For many people, there is a tendency to lose natural lumbar lordosis (curve inward) in this pose. To resist this, engage your **spinal extensors** to lengthen and stabilize your spine. Your **rectus abdominis** stretches slightly.

Thighs

Your **hip flexors** – particularly your **iliopsoas** – engage to maintain hip flexion. Your **quadriceps**, along with your **adductors**, strongly stretch. Although your **hamstrings** initially engage to flex your knee, in this version of the pose try to relax them as much as possible. This is not a major hamstring strengthener.

Spine

Spinal extensors

Rectus abdominis

Psoas major
Iliacus
Hip
Pectineus
Adductor longus
Rectus femoris
Adductor magnus
Sartorius
Vastus medialis
Knee

» CLOSER LOOK

Your one-of-a-kind bone shapes and joint structures determine what your bound angle pose looks like. Some people will never be able to bring their knees to the floor and that is okay. Focus on releasing your hips.

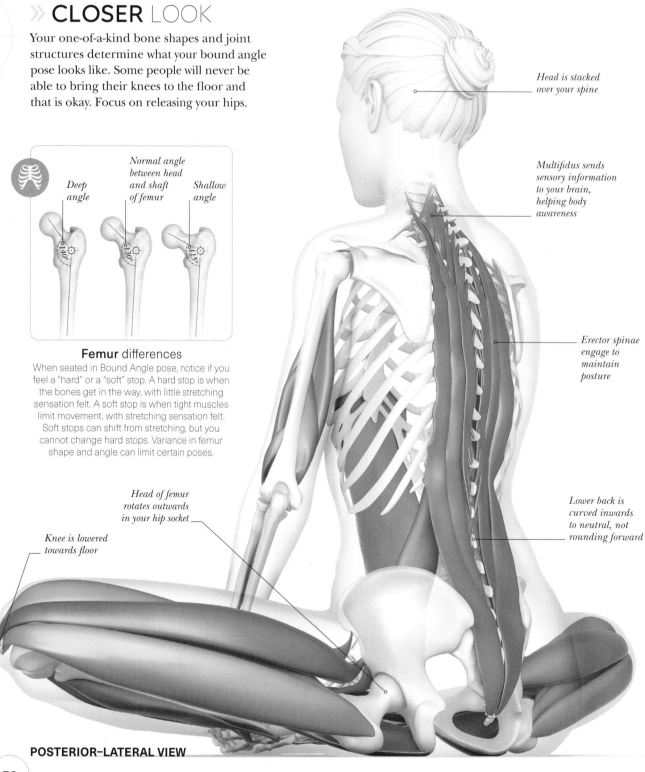

Deep angle

Normal angle between head and shaft of femur

Shallow angle

Femur differences

When seated in Bound Angle pose, notice if you feel a "hard" or a "soft" stop. A hard stop is when the bones get in the way, with little stretching sensation felt. A soft stop is when tight muscles limit movement, with stretching sensation felt. Soft stops can shift from stretching, but you cannot change hard stops. Variance in femur shape and angle can limit certain poses.

Head is stacked over your spine

Multifidus sends sensory information to your brain, helping body awareness

Erector spinae engage to maintain posture

Lower back is curved inwards to neutral, not rounding forward

Head of femur rotates outwards in your hip socket

Knee is lowered towards floor

POSTERIOR–LATERAL VIEW

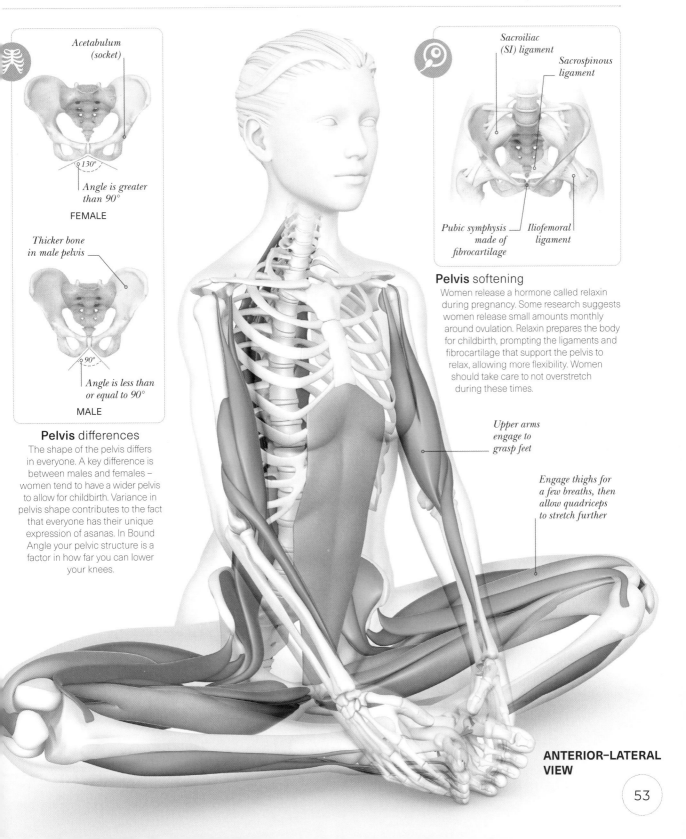

Acetabulum
(socket)

130°

Angle is greater
than 90°

FEMALE

Thicker bone
in male pelvis

90°

Angle is less than
or equal to 90°

MALE

Pelvis differences
The shape of the pelvis differs in everyone. A key difference is between males and females – women tend to have a wider pelvis to allow for childbirth. Variance in pelvis shape contributes to the fact that everyone has their unique expression of asanas. In Bound Angle your pelvic structure is a factor in how far you can lower your knees.

Sacroiliac
(SI) ligament

Sacrospinous
ligament

Pubic symphysis
made of
fibrocartilage

Iliofemoral
ligament

Pelvis softening
Women release a hormone called relaxin during pregnancy. Some research suggests women release small amounts monthly around ovulation. Relaxin prepares the body for childbirth, prompting the ligaments and fibrocartilage that support the pelvis to relax, allowing more flexibility. Women should take care to not overstretch during these times.

Upper arms
engage to
grasp feet

Engage thighs for
a few breaths, then
allow quadriceps
to stretch further

ANTERIOR–LATERAL VIEW

53

CAT
Marjaryasana

This is a gentle kneeling pose that takes the position of a scared cat, warming up joints in your spine, hips, and shoulders. Try exhaling as you move into the pose. This is often done with the next pose, Cow, by flowing from cat to cow with the exhale and inhale.

THE BIG PICTURE

Your back muscles stretch while the muscles on the front of your body – including your chest and abdominal muscles – engage. Muscles in your arms work to stabilize you. Your ribcage is compressed, helping to facilitate a deep exhale into the pose.

ALIGNMENT
Your arms and thighs are fixed in place, with your knees directly under your hips and hands under your shoulders (or slightly forward). The rounding of your spine is as even as possible.

Even curve in spine

Shoulder blades are wide apart

Knees hip distance apart

Hands shoulder distance apart

Fingers spread and palms pressing down

Lower **torso**
Your lumbar spine is in flexion, stretching your **quadratus lumborum**. Your **abdominals** engage to compress your abdomen, squeezing your belly button in towards your spine. Your pelvis is in a posterior pelvic tilt.

Internal obliques

Rectus abdominis

Iliopsoas

Hip

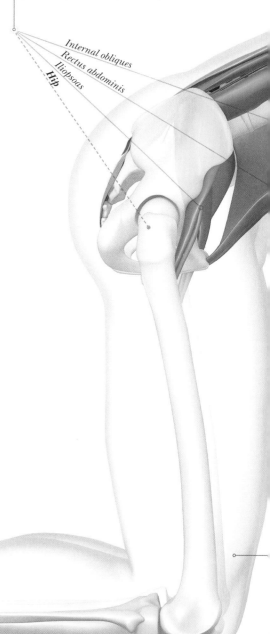

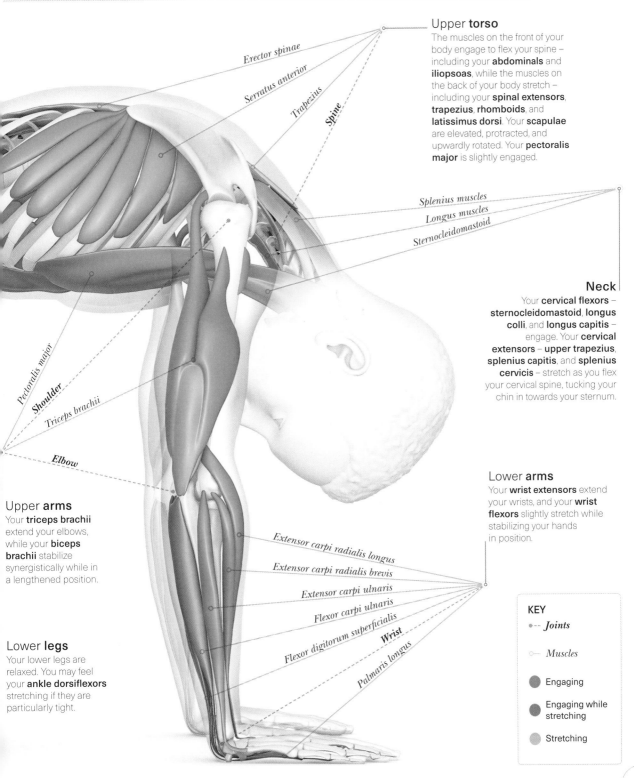

Erector spinae

Serratus anterior

Trapezius

Spine

Upper **torso**
The muscles on the front of your body engage to flex your spine – including your **abdominals** and **iliopsoas**, while the muscles on the back of your body stretch – including your **spinal extensors**, **trapezius**, **rhomboids**, and **latissimus dorsi**. Your **scapulae** are elevated, protracted, and upwardly rotated. Your **pectoralis major** is slightly engaged.

Splenius muscles

Longus muscles

Sternocleidomastoid

Neck
Your **cervical flexors** – **sternocleidomastoid**, **longus colli**, and **longus capitis** – engage. Your **cervical extensors** – **upper trapezius**, **splenius capitis**, and **splenius cervicis** – stretch as you flex your cervical spine, tucking your chin in towards your sternum.

Pectoralis major

Shoulder

Triceps brachii

Elbow

Upper **arms**
Your **triceps brachii** extend your elbows, while your **biceps brachii** stabilize synergistically while in a lengthened position.

Lower **legs**
Your lower legs are relaxed. You may feel your **ankle dorsiflexors** stretching if they are particularly tight.

Extensor carpi radialis longus

Extensor carpi radialis brevis

Extensor carpi ulnaris

Flexor carpi ulnaris

Flexor digitorum superficialis

Wrist

Palmaris longus

Lower **arms**
Your **wrist extensors** extend your wrists, and your **wrist flexors** slightly stretch while stabilizing your hands in position.

KEY
- Joints
- Muscles
- Engaging
- Engaging while stretching
- Stretching

COW
Bitilasana

Mimicking the slightly dipped back of a cow, this gentle kneeling pose incorporates a backbend, and is practised to warm up the spine, hips, and shoulders. Inhale as you enter the pose; you can also alternate between this and Cat pose, in time with your breath.

THE BIG PICTURE

Your abdominal and chest muscles stretch, while your back muscles – including your spinal extensors – engage. Your ribcage is expanding, making it possible to inhale fully. A subtle, even curve is created by the backbend and raised head.

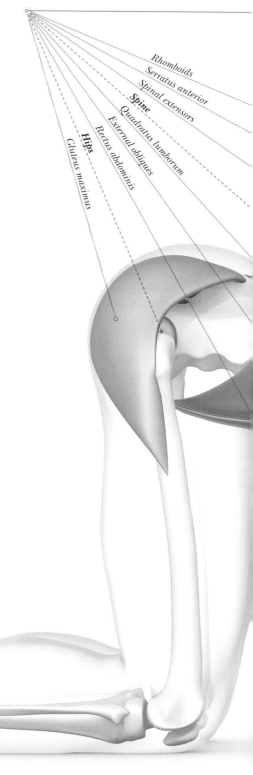

Rhomboids
Serratus anterior
Spinal extensors
Spine
Quadratus lumborum
External obliques
Rectus abdominis
Hips
Gluteus maximus

ALIGNMENT
Your arms and thighs are fixed in place with your knees under your hips and hands under your shoulders (or slightly forward). Your backbend is as even as possible, focusing on lengthening your neck, creating a subtle, even curve.

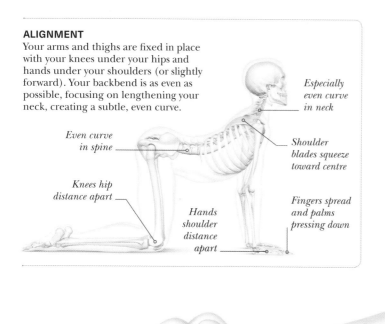

Especially even curve in neck

Even curve in spine

Shoulder blades squeeze toward centre

Knees hip distance apart

Hands shoulder distance apart

Fingers spread and palms pressing down

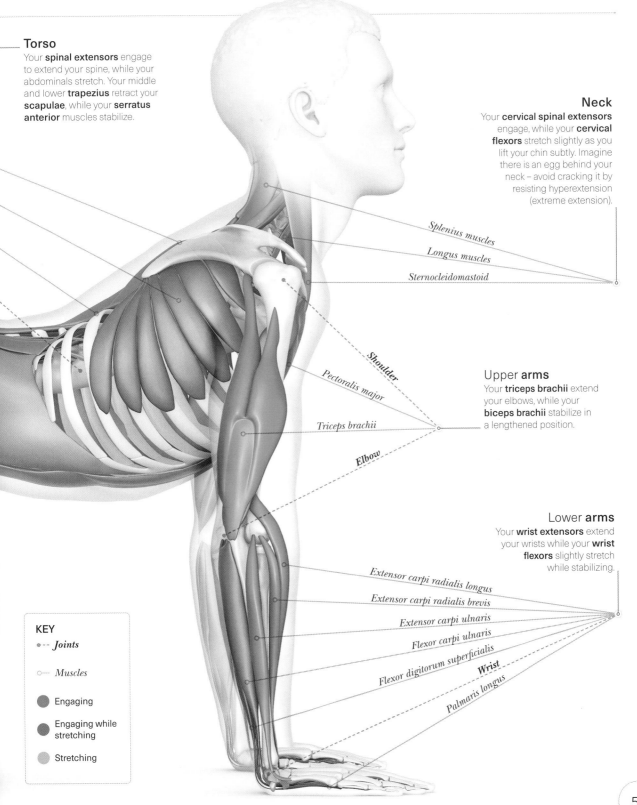

Torso
Your **spinal extensors** engage to extend your spine, while your abdominals stretch. Your middle and lower **trapezius** retract your **scapulae**, while your **serratus anterior** muscles stabilize.

Neck
Your **cervical spinal extensors** engage, while your **cervical flexors** stretch slightly as you lift your chin subtly. Imagine there is an egg behind your neck – avoid cracking it by resisting hyperextension (extreme extension).

Splenius muscles

Longus muscles

Sternocleidomastoid

Shoulder

Pectoralis major

Upper **arms**
Your **triceps brachii** extend your elbows, while your **biceps brachii** stabilize in a lengthened position.

Triceps brachii

Elbow

Lower **arms**
Your **wrist extensors** extend your wrists while your **wrist flexors** slightly stretch while stabilizing.

Extensor carpi radialis longus

Extensor carpi radialis brevis

Extensor carpi ulnaris

Flexor carpi ulnaris

Flexor digitorum superficialis

Wrist

Palmaris longus

KEY
- --- *Joints*
- ○— *Muscles*
- ● Engaging
- ● Engaging while stretching
- ○ Stretching

» CLOSER LOOK

Flowing from the flexion of Cat to the
extension of Cow as you breathe deeply in
and out improves your mind–body connection,
as well as your sense of body awareness.

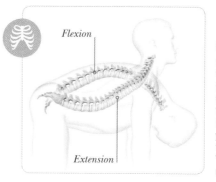

Flexion

Extension

Spine flexion
and **extension**

When your spine flexes, the
front of your body engages while
the back of your body stretches.
When your spine extends, going
into a backbend, the back of
your body engages while the
front of your body stretches.
Your spinal extensors are the
main players in this extension.

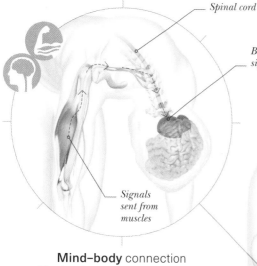

Spinal cord

*Brain receives
signals*

*Signals
sent from
muscles*

*Trapezius
stretches*

Mind–body connection

We often think of the brain as controlling
our muscles. This is true: those motor
signals tell your muscles what to do.
However, your nervous system is a two-way
conversation. Your body sends tons of
sensory signals to your brain. Yoga improves
mind–body connection by encouraging
you to listen to your body.

*Ankles and feet
are relaxed (in
plantar flexion)*

*Middle finger is
facing forwards*

**ANTERIOR–LATERAL
VIEW OF CAT**

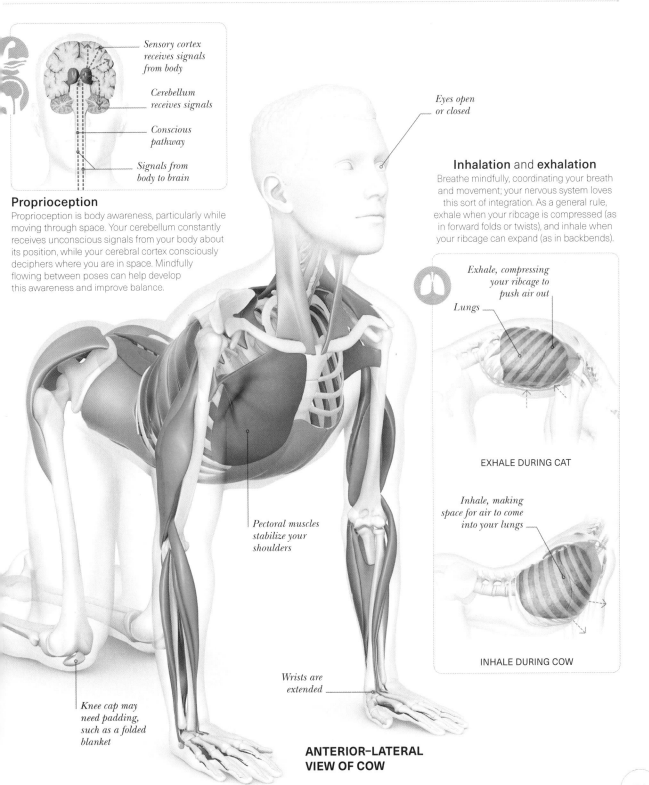

Sensory cortex receives signals from body

Cerebellum receives signals

Conscious pathway

Signals from body to brain

Proprioception

Proprioception is body awareness, particularly while moving through space. Your cerebellum constantly receives unconscious signals from your body about its position, while your cerebral cortex consciously deciphers where you are in space. Mindfully flowing between poses can help develop this awareness and improve balance.

Eyes open or closed

Inhalation and exhalation

Breathe mindfully, coordinating your breath and movement; your nervous system loves this sort of integration. As a general rule, exhale when your ribcage is compressed (as in forward folds or twists), and inhale when your ribcage can expand (as in backbends).

Exhale, compressing your ribcage to push air out

Lungs

EXHALE DURING CAT

Inhale, making space for air to come into your lungs

INHALE DURING COW

Pectoral muscles stabilize your shoulders

Knee cap may need padding, such as a folded blanket

Wrists are extended

ANTERIOR–LATERAL VIEW OF COW

59

COW FACE
Gomukhasana

This seated pose involves unique actions of your shoulder joints. This can be helpful in stretching out tight shoulders, particularly if you work at a desk and spend a lot of time typing – but you should avoid this pose if you have a rotator cuff injury. Switch arms and notice if you feel a difference between each side.

THE BIG PICTURE

In this seated pose, you particularly stretch around your shoulders and the outside of your hips and buttocks. You are also engaging key postural muscles to counteract slouching or rounding forward.

KEY

•-- *Joints*

○— *Muscles*

● Engaging

● Engaging while stretching

● Stretching

Top **arm**
Your shoulder flexors – **anterior deltoid** and **pectoralis major** – flex your shoulder. Your **middle deltoid** and **supraspinatus** stabilize and abduct it, and your **infraspinatus**, **teres minor**, and **posterior deltoid** engage to externally rotate. Your elbow flexors engage and **triceps brachii** stretches.

Torso
Your **spinal extensors** and **transversus abdominis** engage to slightly extend and stabilize your spine, while your **rectus abdominis** stretches. Your **rhomboids** engage to retract your **scapulae**.

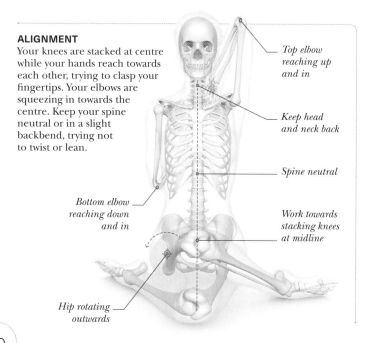

ALIGNMENT
Your knees are stacked at centre while your hands reach towards each other, trying to clasp your fingertips. Your elbows are squeezing in towards the centre. Keep your spine neutral or in a slight backbend, trying not to twist or lean.

Top elbow reaching up and in

Keep head and neck back

Spine neutral

Bottom elbow reaching down and in

Work towards stacking knees at midline

Hip rotating outwards

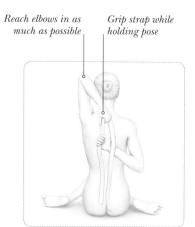

Reach elbows in as much as possible

Grip strap while holding pose

VARIATION
If you cannot reach your hands together, use a strap or towel to extend your reach. If you hold for approximately 10 breaths, you may find you can walk your fingers in closer towards each other.

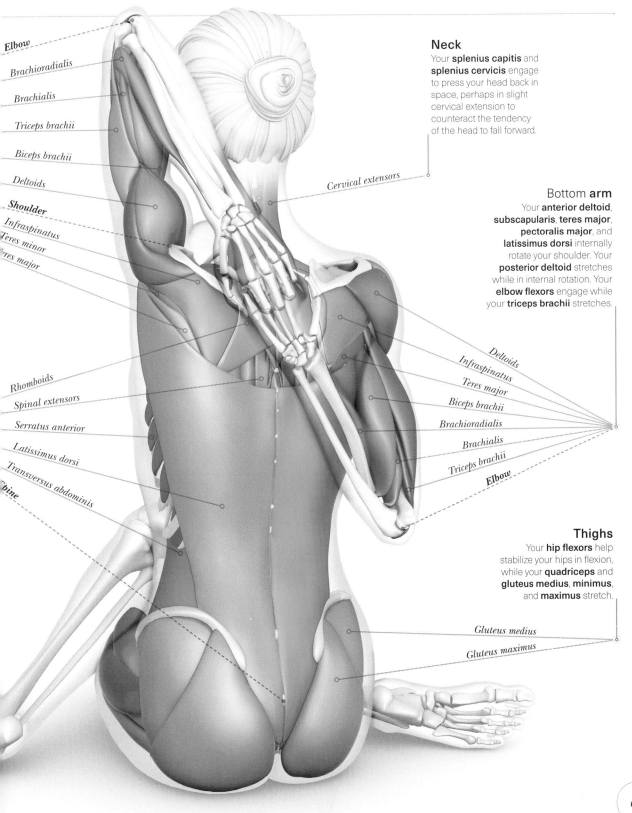

Elbow

Brachioradialis

Brachialis

Triceps brachii

Biceps brachii

Deltoids

Shoulder

Infraspinatus

Teres minor

Teres major

Rhomboids

Spinal extensors

Serratus anterior

Latissimus dorsi

Transversus abdominis

Spine

Neck
Your **splenius capitis** and **splenius cervicis** engage to press your head back in space, perhaps in slight cervical extension to counteract the tendency of the head to fall forward.

Cervical extensors

Bottom **arm**
Your **anterior deltoid**, **subscapularis**, **teres major**, **pectoralis major**, and **latissimus dorsi** internally rotate your shoulder. Your **posterior deltoid** stretches while in internal rotation. Your **elbow flexors** engage while your **triceps brachii** stretches.

Deltoids

Infraspinatus

Teres major

Biceps brachii

Brachioradialis

Brachialis

Triceps brachii

Elbow

Thighs
Your **hip flexors** help stabilize your hips in flexion, while your **quadriceps** and **gluteus medius, minimus,** and **maximus** stretch.

Gluteus medius

Gluteus maximus

» **CLOSER** LOOK

Cow Face works your shoulders dynamically – including your deltoid muscles. Compression at your shoulder joint can also lead to cardiovascular shifts both in local blood vessels and system-wide.

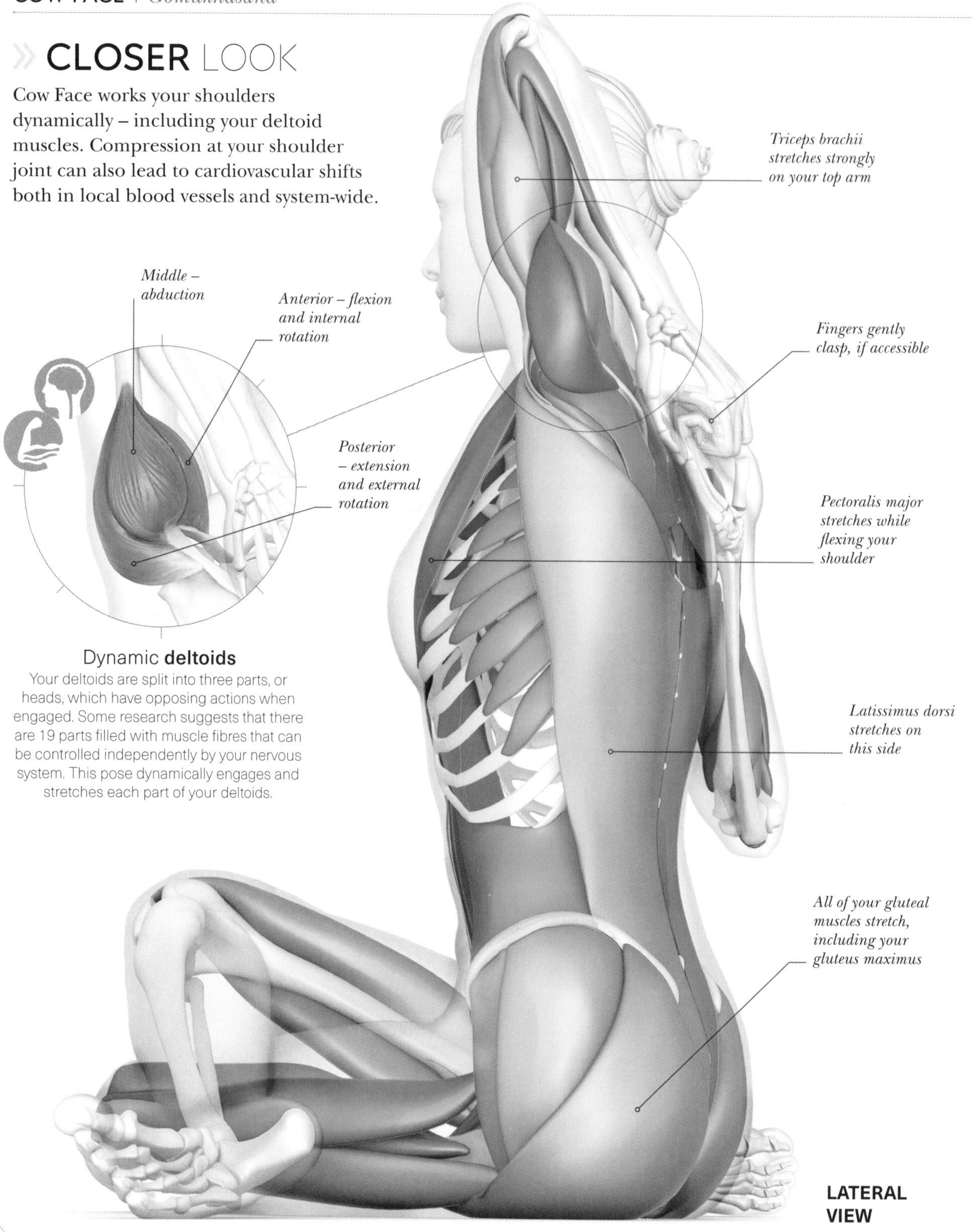

Middle – abduction

Anterior – flexion and internal rotation

Posterior – extension and external rotation

Dynamic **deltoids**

Your deltoids are split into three parts, or heads, which have opposing actions when engaged. Some research suggests that there are 19 parts filled with muscle fibres that can be controlled independently by your nervous system. This pose dynamically engages and stretches each part of your deltoids.

Triceps brachii stretches strongly on your top arm

Fingers gently clasp, if accessible

Pectoralis major stretches while flexing your shoulder

Latissimus dorsi stretches on this side

All of your gluteal muscles stretch, including your gluteus maximus

LATERAL VIEW

*Eyes open
or closed*

*Collar bones move
slightly with
your arms*

*Middle deltoid
stretches as you
slightly squeeze
your elbow
inwards*

*Pectoralis major
engages to
strongly adduct
your shoulder*

*Psoas major on
both sides engage
to flex your hips*

*Ankles and
feet are relaxed*

**ANTERIOR
VIEW**

Flexion

*External
rotation*

Extension

*Internal
rotation*

Range of movement

Your body has the potential to do many actions,
but our modern-day lifestyle limits its opportunities.
Humans are designed to go through more joint
actions more regularly. Your yoga practice helps you
maintain these capabilities in full range of motion
(ROM). When it comes to ROM, if you don't
use it, you lose it.

*Vasoconstriction
(vessel constricting)*

*Vasodilation
(vessel opening)*

Blood vessel changes

There is slight pressure on your blood vessels at
your shoulders, similar to a loose tourniquet. When
you release the pose, blood rushes to the area. This
vascular pressure causes an increase of nitric oxide
(NO), encouraging blood vessel dilation, slightly
lowering blood pressure and increasing relaxation.

63

SIDE BEND

Parivrtta Janu Sirsasana

This seated, lateral side stretch allows you to mobilize your spine in a way that you probably don't often move it in everyday life. The novel movement that this pose involves benefits your intervertebral discs, nervous system, and fascia.

THE BIG PICTURE

As you bend deeply to the side, muscles along your spine stretch and strengthen. Your shoulder muscles engage to reach your arms over your head, and your thigh muscles on both sides stretch in different ways.

Neck

To rotate your neck, your **rotatores**, **multifidus**, **sternocleidomastoid**, and **semispinalis cervicis** engage on the side towards the ground (model's right in this image), while stretching on the upward-facing side. Your **splenius capitis** and **splenius cervicis** engage on the upward-facing side (model's left in this image), while stretching on the downward-facing side.

Sternocleidomastoid

Brachioradialis
Serratus anterior
Shoulder
Deltoids
Triceps brachii
Biceps brachii
Brachialis
Elbow

Arms

Your **shoulder flexors** – including your **anterior deltoids** – engage. Your **middle deltoids** and **supraspinatus** engage to abduct your shoulders, and they are externally rotated by your **posterior deltoids**, **infraspinatus**, and **teres minor muscles**. Your **brachialis**, **biceps**, and **brachioradialis** muscles flex your elbows.

KEY

- •-- *Joints*
- ○-- *Muscles*
- ● Engaging
- ● Engaging while stretching
- ● Stretching

Extended lower leg

Your **ankle dorsiflexors** engage to dorsiflex your ankle and extend your toes. If you are grabbing your foot and pulling, you probably feel a stretch in your **calf muscles**, along with your **plantar muscles** and **fascia**.

Plantar fascia
Tibialis anterior
Flexor d. longus
Gastrocnemius
Ankle

Thigh

Your **hamstrings** and **gluteus maximus** stretch, while your **quadriceps** engage to extend your knee. Also, your **internal rotators** – including **gluteus medius**, **gluteus minimus**, and **tensor fasciae latae** – engage while lengthening. You may feel a stretch in your **iliotibial band**.

Torso

On the side towards the ground your **external abdominal obliques**, **erector spinae**, and **quadratus lumborum** engage while the upper side stretches, to laterally flex your spine. On both sides, your **rotatores** and **multifidus** rotate your spine and send signals to your brain about where your spine is in space. Your **transversus abdominis** engages to stabilize your spine.

Spine

Spinal extensors

External obliques

ALIGNMENT

Avoid rounding forward by reaching your top shoulder blade back, as if you are trying to press towards an imaginary wall. Focus on finding length in your spine and broadness in your chest.

Turn head comfortably

Spine elongating

Bring shoulder blade back

Knees soft, not locked

Chest broad

Flexed leg

Your **adductors**, **quadriceps**, and **iliopsoas** stretch. Although your **hamstrings** engage initially to flex your knee in place, try to relax your leg muscles while holding the pose.

Gluteus maximus

Pectineus

Psoas major

Adductor magnus

Sartorius

Vastus medialis

Knee

Rectus femoris

Vastus lateralis

Semitendinosus

Rectus femoris

Iliopsoas

Adductor magnus

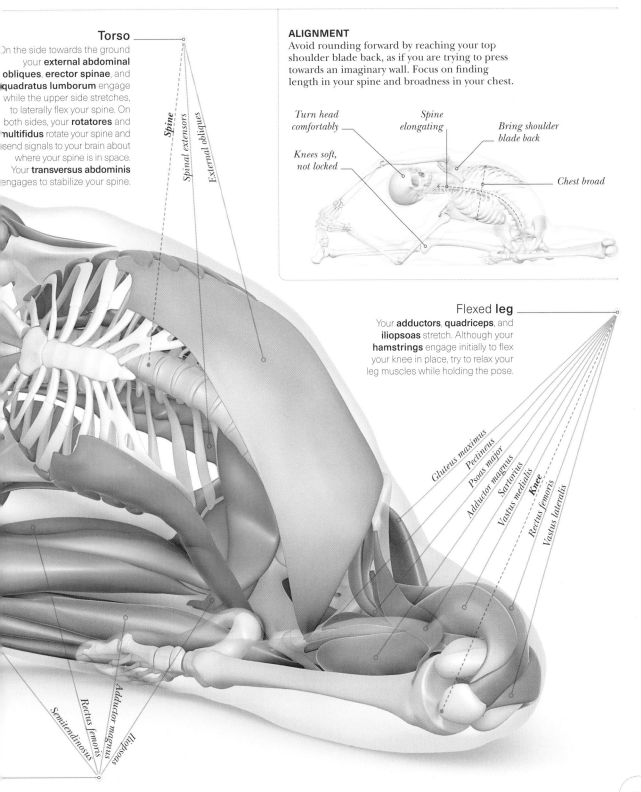

» CLOSER LOOK

Seated side bend is a one-sided movement that dynamically affects your abdominals, back muscles, and spinal discs. You don't have to be able to reach your foot with either hand to do this pose; your arms can simply reach to the side.

Disc health

When side-bending (lateral flexion of the spine), your intervertebral discs push to the sides. As you bend to the right, your discs shift to the left (and vice versa). The cartilage in your spine allows for this natural process.

Vertebra ___

Disc pushes to opposite side of bend

Rectus abdominis ___

External oblique ___

Internal oblique ___

Transversus abdominis ___

Abdominal structure

Your criss-crossing abdominal muscles provide multi-layered support for your internal organs and allow your torso to move. Legend has it that in 1888, Dr. Dunlop, a surgeon, was watching his son on his tricycle bouncing due to the poor design of the wheels, causing a headache. Inspired by the structure of the abdominals, he designed a tyre for a smoother ride and fewer flats.

Latissimus dorsi stretches on the upper side ___

Quadratus lumborum stretches on upper side

Quadratus lumborum engages on lower side

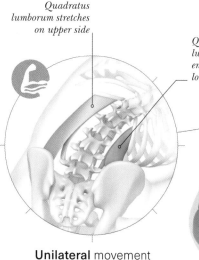

Unilateral movement

Your quadratus lumborum (QL) is important for holding posture. When the erector spinae are weak it picks up the slack. Keeping your spine erect is a big job for this little muscle, leading to muscle fatigue and even pain. This pose helps by stretching and engaging the QL.

Pressure and balance
Notice and feel the point of contact of your body on the floor in this pose. It is a little different for everybody. Notice how the pressure points shift as you transition in and out of the pose.

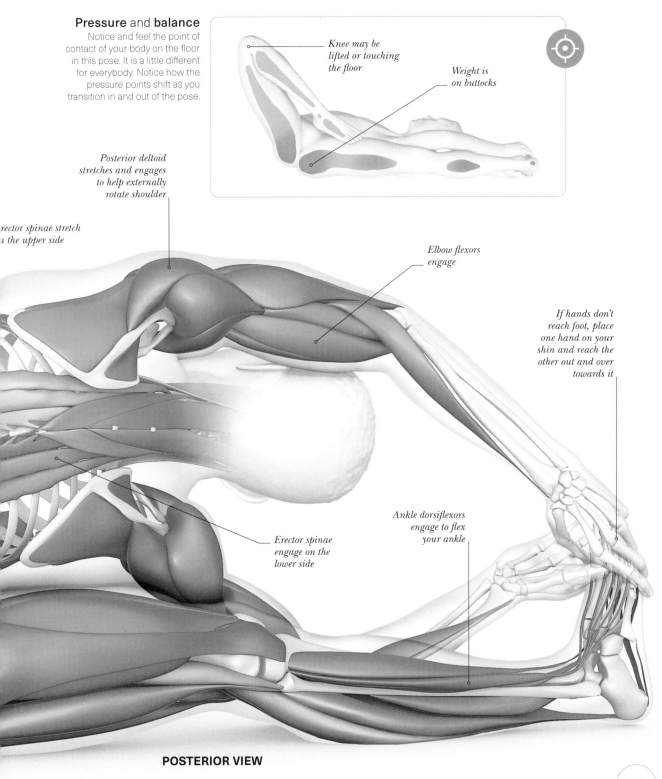

Knee may be lifted or touching the floor

Weight is on buttocks

Posterior deltoid stretches and engages to help externally rotate shoulder

Erector spinae stretch *on the upper side*

Elbow flexors engage

If hands don't reach foot, place one hand on your shin and reach the other out and over towards it

Erector spinae engage on the lower side

Ankle dorsiflexors engage to flex your ankle

POSTERIOR VIEW

67

SEATED TWIST

Ardha Matsyendrasana

This seated twist will wake up small muscles along your spine and stimulate digestion. Practising twists mindfully in yoga can help prevent injury from twists you do in everyday life. Take care not to twist too far if you have spinal disc issues or osteoporosis.

THE BIG PICTURE

Your back muscles and abdominals dynamically engage and stretch as you rotate your spine. Your thighs and hips – particularly around your buttocks – are stretching as they rotate outwards. Your lowered arm presses down to help you find more length along your spine.

KEY

- •-- *Joints*
- ○— *Muscles*
- ● Engaging
- ● Engaging while stretching
- ● Stretching

ALIGNMENT
Prioritize elongating your spine over rotating more or leaning. If you do decide to rotate more deeply, try to use your core muscles instead of pulling with the external force of your arms.

Spine elongating

Keep the rotation as even as possible

Pelvis rotates slightly with you

Arm presses down

Neck
To rotate your neck, on the contralateral side of axial rotation (side you are rotating away from, model's left side on this image), your **rotatores**, **multifidus**, **sternocleidomastoid**, and **semispinalis cervicis** engage while stretching on the ipsilateral side (the side you are rotating towards). Your **splenius capitis** and **splenius cervicis** engage on the ipsilateral side, and stretch on the contralateral.

Sternocleidomastoid

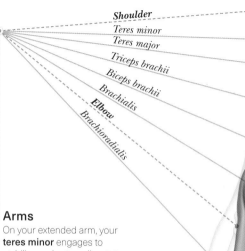

Shoulder
Teres minor
Teres major
Triceps brachii
Biceps brachii
Brachialis
Elbow
Brachioradialis

Arms
On your extended arm, your **teres minor** engages to stabilize and externally rotate your shoulder, while your **teres major** extends your shoulder. Your **elbow flexors** and **triceps** are dynamically engaging to help hold your arm in place, pushing down into the ground to help elongate your spine. On your flexed arm, your **elbow flexors** engage while your **triceps** stretches slightly.

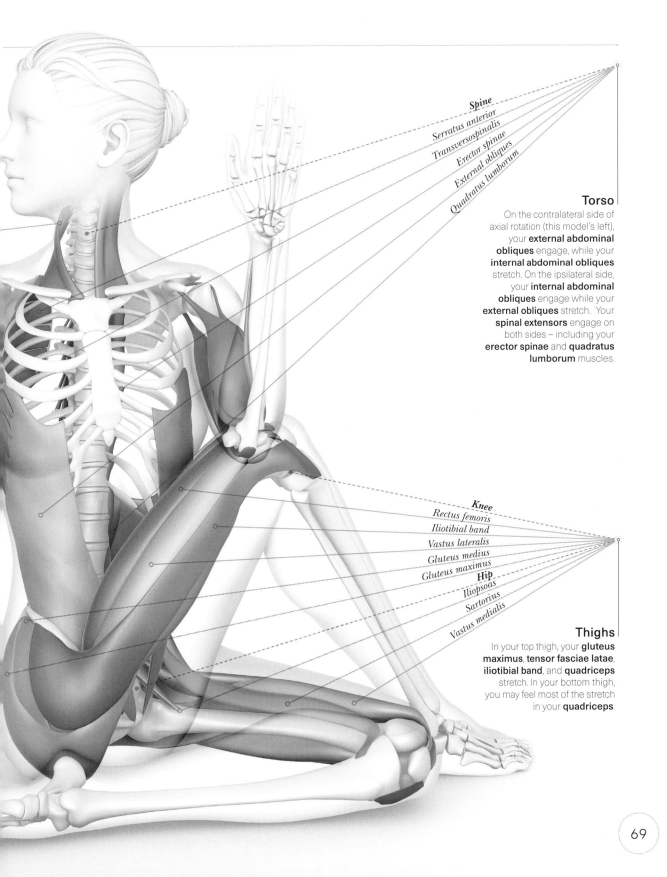

Spine
Serratus anterior
Transversospinalis
Erector spinae
External obliques
Quadratus lumborum

Torso

On the contralateral side of axial rotation (this model's left), your **external abdominal obliques** engage, while your **internal abdominal obliques** stretch. On the ipsilateral side, your **internal abdominal obliques** engage while your **external obliques** stretch. Your **spinal extensors** engage on both sides – including your **erector spinae** and **quadratus lumborum** muscles.

Knee
Rectus femoris
Iliotibial band
Vastus lateralis
Gluteus medius
Gluteus maximus
Hip
Iliopsoas
Sartorius
Vastus medialis

Thighs

In your top thigh, your **gluteus maximus, tensor fasciae latae, iliotibial band**, and **quadriceps** stretch. In your bottom thigh, you may feel most of the stretch in your **quadriceps**.

» CLOSER LOOK

Spinal twists affect the discs between your vertebrae
and your sacroiliac joint. Although this action may
not "wring out toxins" as is sometimes claimed,
it does encourage healthy digestive movement in
your intestines, known as peristalsis.

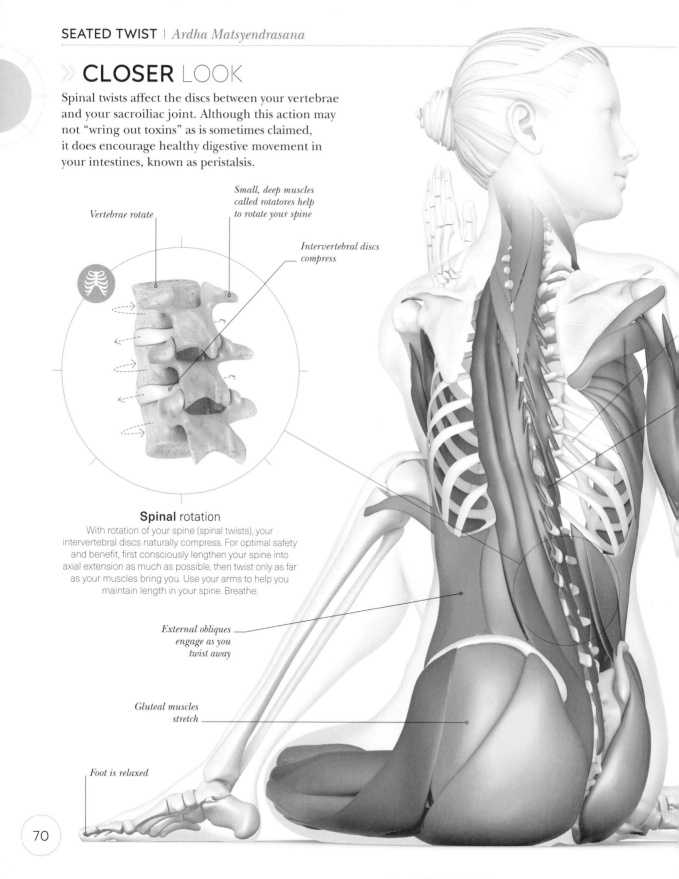

Vertebrae rotate

*Small, deep muscles
called rotatores help
to rotate your spine*

*Intervertebral discs
compress*

Spinal rotation

With rotation of your spine (spinal twists), your
intervertebral discs naturally compress. For optimal safety
and benefit, first consciously lengthen your spine into
axial extension as much as possible, then twist only as far
as your muscles bring you. Use your arms to help you
maintain length in your spine. Breathe.

*External obliques
engage as you
twist away*

*Gluteal muscles
stretch*

Foot is relaxed

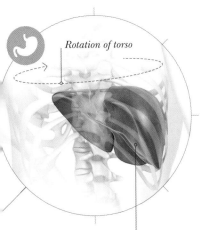

Rotation of torso

Wringing out **toxins**

You may have heard that spinal twists "wring out toxins". However, your liver efficiently deals with toxins automatically. Although mechanically compressing the organs may be beneficial, evidence does not show that this contributes to "detoxification". You can instead visualize wringing out negative energy as you twist, for psychological benefit.

Your liver naturally detoxifies

Deep to the erector spinae, the multifidus is dynamically engaging

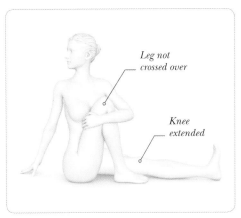

Leg not crossed over

Knee extended

VARIATION
For a gentler twist, keep one leg extended and consider not crossing the lifted leg over the midline. Use your arm wrapped around the leg to sit tall as you twist.

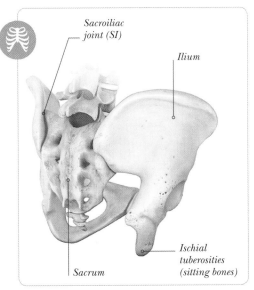

Sacroiliac joint (SI)

Ilium

Ischial tuberosities (sitting bones)

Sacrum

Sacroiliac joint

Allow your sitting bones to move slightly on the floor with the twist. If you anchor them down, the twist puts a lot of pressure on the structure of the SI joint, which can cause aches. Alternatively, allowing too much movement in your SI joint can also lead to achiness. Find the middle way for your body.

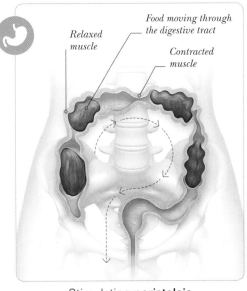

Food moving through the digestive tract

Relaxed muscle

Contracted muscle

Stimulating **peristalsis**

In your digestive tract, peristalsis is the involuntary engagement of smooth muscles to move digesting food (see p.39). Thankfully, you don't have to consciously tell your stomach to empty into the small intestines. Stress and a sedentary lifestyle can affect peristalsis, causing digestive issues. Twisting can stimulate healthy peristalsis.

POSTERIOR–LATERAL VIEW

CHILD'S POSE
Balasana

Reminiscent of the fetal position and with your weight supported by the floor, the restorative forward bend of child's pose can be a deeply relaxing, restful posture for many. It provides a deep but gentle stretch for your back muscles, calming both body and mind.

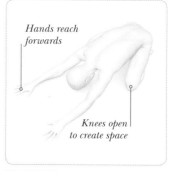

Hands reach forwards

Knees open to create space

VARIATION
Another option is to separate your knees and bring your hands forwards. This allows more space for the torso and is a common resting pose during sequences such as Sun salutations.

THE BIG PICTURE

With as little muscle engagement as possible, your body releases down. In particular your back muscles, buttocks, and ankles stretch out. As you breathe deeply, the muscles in and around your ribcage dynamically engage and stretch with each breath.

Neck and upper **arms**
Your neck muscles are passive with your **splenius capitis** and **splenius cervicis** stretching. Your **posterior deltoids** slightly stretch while your shoulders are in internal rotation. Your **arm muscles** are passive with your forearms pronated, allowing the backs of your hands to rest on the floor.

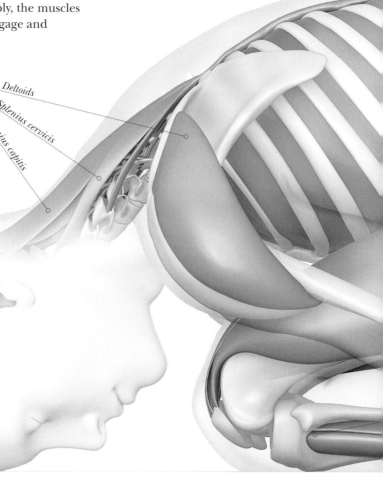

Deltoids

Splenius cervicis

Splenius capitis

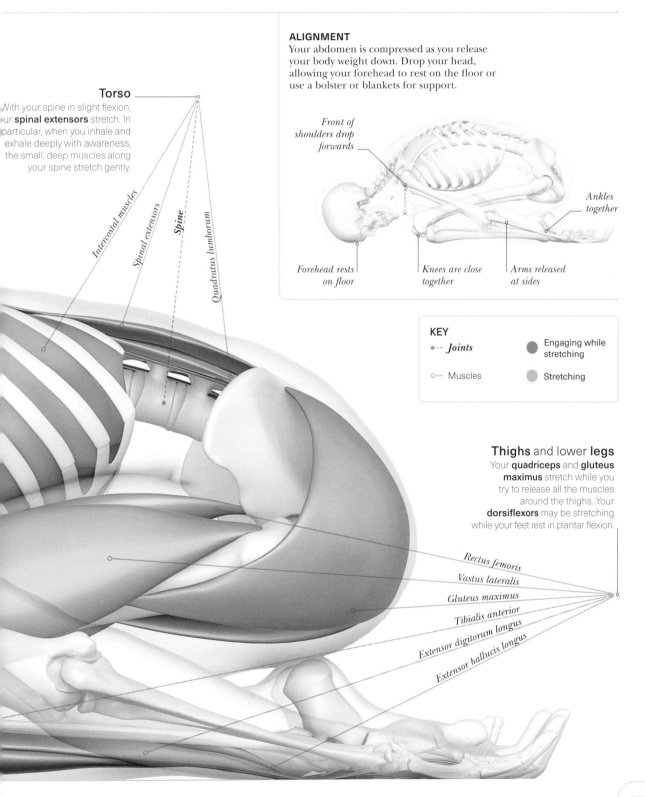

Torso

With your spine in slight flexion, your **spinal extensors** stretch. In particular, when you inhale and exhale deeply with awareness, the small, deep muscles along your spine stretch gently.

Intercostal muscles

Spinal extensors

Spine

Quadratus lumborum

ALIGNMENT

Your abdomen is compressed as you release your body weight down. Drop your head, allowing your forehead to rest on the floor or use a bolster or blankets for support.

Front of shoulders drop forwards

Ankles together

Forehead rests on floor

Knees are close together

Arms released at sides

KEY

•--- *Joints*

○— Muscles

● Engaging while stretching

● Stretching

Thighs and lower legs

Your **quadriceps** and **gluteus maximus** stretch while you try to release all the muscles around the thighs. Your **dorsiflexors** may be stretching while your feet rest in plantar flexion.

Rectus femoris

Vastus lateralis

Gluteus maximus

Tibialis anterior

Extensor digitorum longus

Extensor hallucis longus

» CLOSER LOOK

Child's Pose can be an opportunity to rest, take deep breaths, relax tired muscles, and access a primal sense of safety. If comfortable, you can use this pose as a place of rest and rejuvenation between challenging poses.

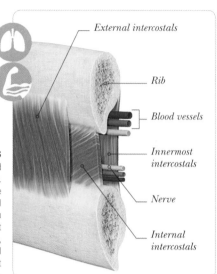

External intercostals

Rib

Blood vessels

Innermost intercostals

Nerve

Internal intercostals

Splenius muscles

If needed, put a block or cushion beneath your forehead

Intercostal muscles

Your intercostals are criss-crossed and layered, like your abdominals. Your external intercostals engage to help you inhale. Your internal intercostals engage to help you forcefully exhale. Your innermost intercostals stabilize your ribs, stretching when you inhale. Feel how dynamic your rib movement is while you take deep breaths here.

Head rest

Throughout the day, your neck muscles have the job of holding up the 5kg (11lb) bowling ball that is your head. This muscular activity keeps your nervous system on slight alert. Letting the muscles of your neck and head completely relax lets your nervous system know it is safe to rest.

Neck muscles are completely relaxed

Quadriceps stretch

SUPERIOR-LATERAL VIEW

Fetal position

This pose may evoke comfort as it is reminiscent of being in your mother's womb. In the fetal position, most of your joints are in flexion, protecting your abdominal organs from harm. Notice how your body moves with each breath: your torso rising and broadening with each inhale and releasing back as you exhale.

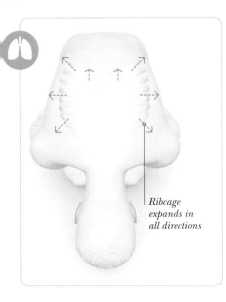

Ribcage expands in all directions

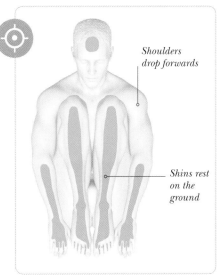

Shoulders drop forwards

Shins rest on the ground

Pressure points

Allow your body to release down completely, with your shins, feet, forearms, hands, and forehead all resting on the ground. If your body doesn't make this shape, use blankets and props to find support.

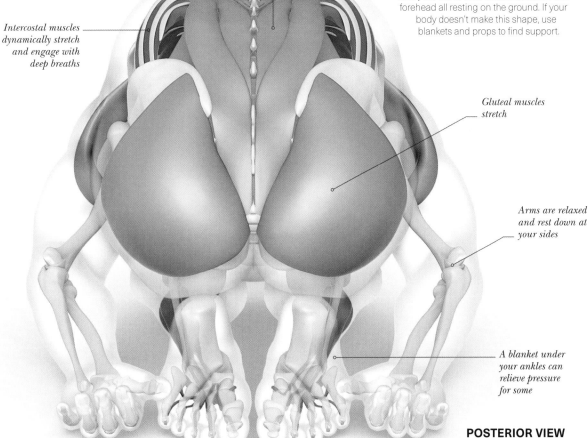

Spinal extensors stretch while in relaxed flexion

Intercostal muscles dynamically stretch and engage with deep breaths

Gluteal muscles stretch

Arms are relaxed and rest down at your sides

A blanket under your ankles can relieve pressure for some

POSTERIOR VIEW

CAMEL
Ustrasana

Camel is an energetic backbend that can leave you feeling confident, ready to take on the day. This pose counteracts our flexion-driven modern lifestyles by broadening the chest. It is challenging but can be adapted for anyone who struggles to reach their feet.

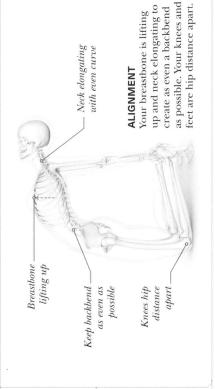

Breastbone lifting up

Keep backbend as even as possible

Knees hip distance apart

Neck elongating with even curve

ALIGNMENT
Your breastbone is lifting up and neck elongating to create as even a backbend as possible. Your knees and feet are hip distance apart.

THE BIG PICTURE

The front of your body – including your abdominals and thighs – stretches, while the back of your body – including your back muscles, buttocks, and thighs – engages. You may also feel stretching on the soles of your feet as you tuck your toes under.

Neck
Your **cervical extensors** engage to extend your neck, while your **cervical flexors** stabilize, preventing your head from being thrown back, and creating an even, controlled curve.

Upper trapezius
Splenius muscles
Longus muscles
Sternocleidomastoid

Torso
Your **spinal extensors** engage to extend your spine while your **abdominals** stretch. Your **pectorals** stretch as you broaden your chest. Your **middle** and **lower trapezius** work with your **rhomboids** to retract and stabilize your scapulae, while your **serratus anterior** stretches.

Pectoralis major
Serratus anterior
Erector spinae
Spine
Quadratus lumborum
Rectus abdominis

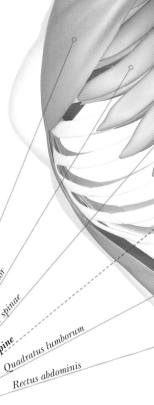

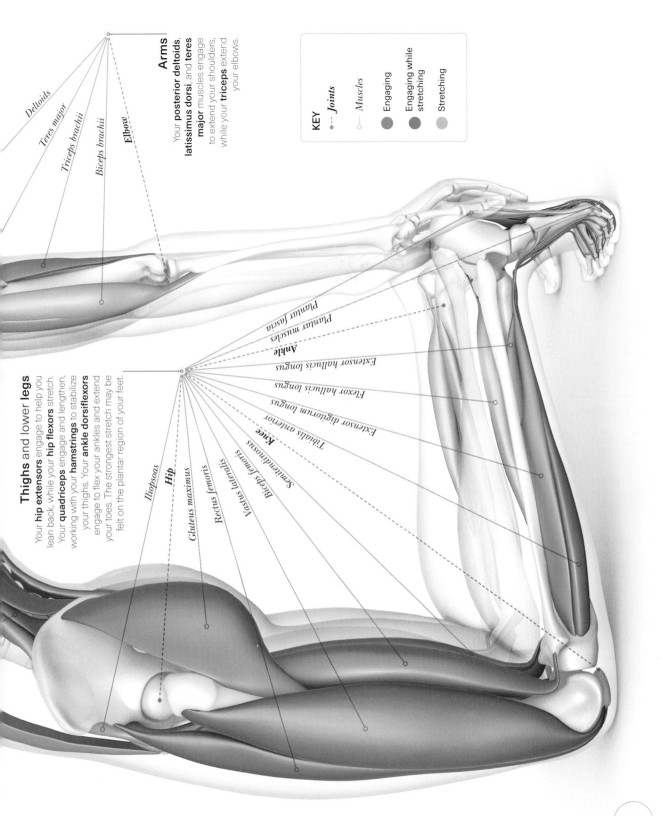

Arms

Your **posterior deltoids**, **latissimus dorsi**, and **teres major** muscles engage to extend your shoulders, while your **triceps** extend your elbows.

Deltoids

Teres major

Triceps brachii

Biceps brachii

Elbow

Plantar fascia

Plantar muscles

Ankle

Extensor hallucis longus

Flexor hallucis longus

Extensor digitorum longus

Tibialis anterior

Knee

Semitendinosus

Biceps femoris

Vastus lateralis

Rectus femoris

Gluteus maximus

Hip

Iliopsoas

Thighs and lower legs

Your **hip extensors** engage to help you lean back, while your **hip flexors** stretch. Your **quadriceps** engage and lengthen, working with your **hamstrings** to stabilize your thighs. Your **ankle dorsiflexors** engage to flex your ankles and extend your toes. The strongest stretch may be felt on the plantar region of your feet.

» CLOSER LOOK

Camel can be great for your spinal disc health and posture. However, make sure you warm up first, and take care with the position of your neck.

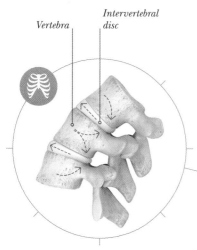

Vertebra

Intervertebral disc

Spinal extension

Backbends (spinal extension) push your intervertebral discs slightly forwards, while strengthening your back muscles. This is great for the health of your discs, and can be applied therapeutically for disc issues. Consult your healthcare team and a qualified yoga professional about your unique conditions first.

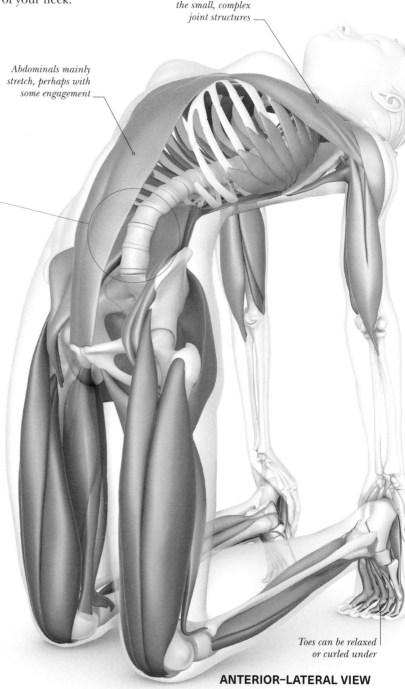

This subtle, controlled neck extension protects the small, complex joint structures

Abdominals mainly stretch, perhaps with some engagement

Toes can be relaxed or curled under

ANTERIOR–LATERAL VIEW

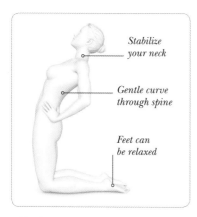

Stabilize your neck

Gentle curve through spine

Feet can be relaxed

VARIATION

For a gentler backbend, press your hands into your hips as you lean back slightly into the pose. You could also bring your hands to blocks placed alongside your shins.

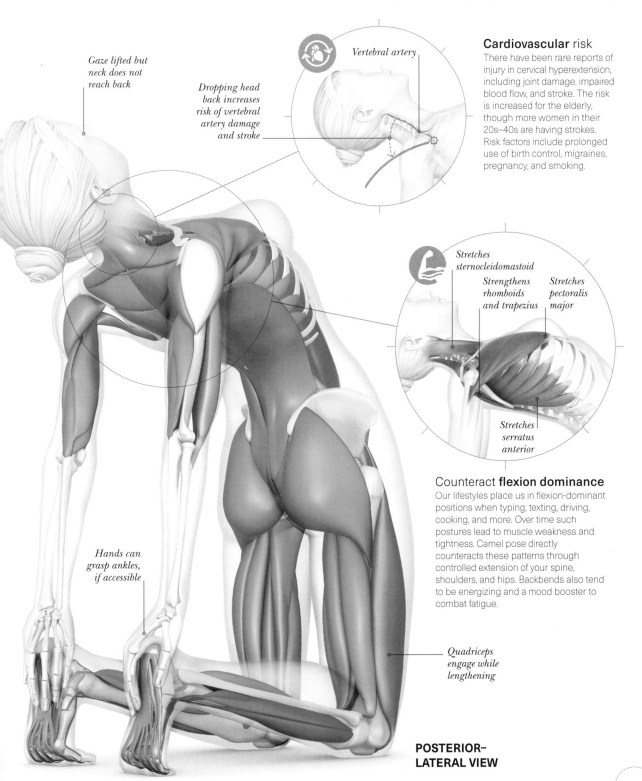

Gaze lifted but neck does not reach back

Dropping head back increases risk of vertebral artery damage and stroke

Vertebral artery

Cardiovascular risk
There have been rare reports of injury in cervical hyperextension, including joint damage, impaired blood flow, and stroke. The risk is increased for the elderly, though more women in their 20s–40s are having strokes. Risk factors include prolonged use of birth control, migraines, pregnancy, and smoking.

Stretches sternocleidomastoid

Strengthens rhomboids and trapezius

Stretches pectoralis major

Stretches serratus anterior

Counteract flexion dominance
Our lifestyles place us in flexion-dominant positions when typing, texting, driving, cooking, and more. Over time such postures lead to muscle weakness and tightness. Camel pose directly counteracts these patterns through controlled extension of your spine, shoulders, and hips. Backbends also tend to be energizing and a mood booster to combat fatigue.

Hands can grasp ankles, if accessible

Quadriceps engage while lengthening

POSTERIOR- LATERAL VIEW

79

KING PIGEON
Eka Pada Rajakapotasana

Pigeon pose, as practised today, is not a traditional yoga asana. This modern kneeling backbend can be modified to offer therapeutic benefits for sciatica and back pain, with suitable options for everyone. Make sure you are warmed up and move slowly into this pose.

THE BIG PICTURE

This version of the pose deeply stretches your hips, buttocks, thighs, abdomen, chest, and shoulders. Muscles in your arms, back, and hips engage to hold you in the pose, preventing you from toppling over.

KEY

●-- *Joints*

○— *Muscles*

● Engaging

● Engaging while stretching

● Stretching

Arms
Your **shoulder flexors** engage. Your **deltoids** dynamically engage to bring you into the position and then help pull your leg in. Your **brachialis**, **biceps**, and **brachioradialis** engage to flex your elbow, while your **triceps** stretch.

Brachioradialis
Brachialis
Triceps brachii
Biceps brachii
Deltoids
Shoulder

Back **thigh**
Your **hip extensors** are working to extend your hip, while your **quadriceps** maintain knee extension. Your **hip flexors** stretch strongly.

Gluteus maximus
Tensor fasciae latae
Biceps femoris
Semitendinosus
Rectus femoris
Knee

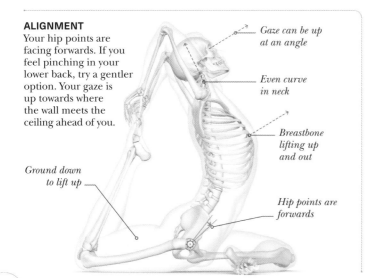

ALIGNMENT
Your hip points are facing forwards. If you feel pinching in your lower back, try a gentler option. Your gaze is up towards where the wall meets the ceiling ahead of you.

Gaze can be up at an angle

Even curve in neck

Breastbone lifting up and out

Ground down to lift up

Hip points are forwards

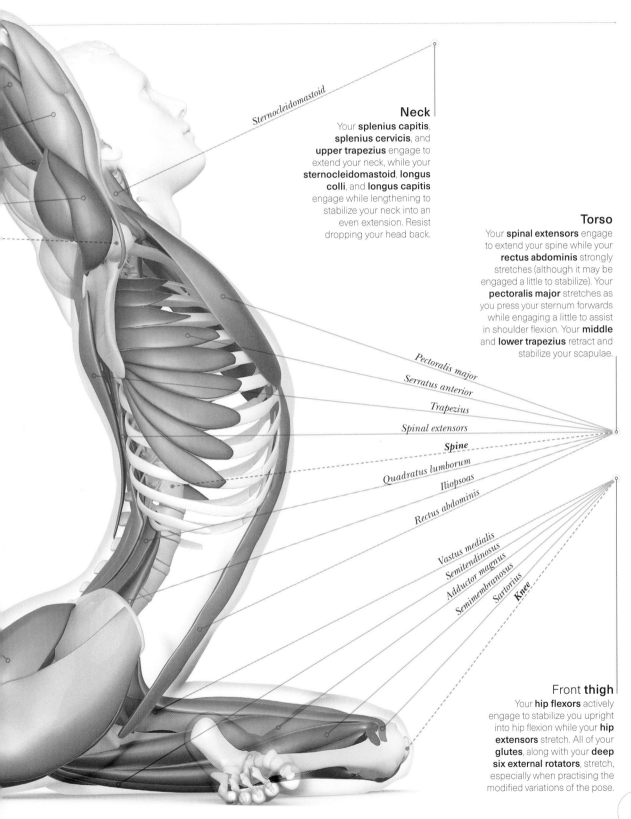

Neck

Your **splenius capitis**, **splenius cervicis**, and **upper trapezius** engage to extend your neck, while your **sternocleidomastoid**, **longus colli**, and **longus capitis** engage while lengthening to stabilize your neck into an even extension. Resist dropping your head back.

Torso

Your **spinal extensors** engage to extend your spine while your **rectus abdominis** strongly stretches (although it may be engaged a little to stabilize). Your **pectoralis major** stretches as you press your sternum forwards while engaging a little to assist in shoulder flexion. Your **middle** and **lower trapezius** retract and stabilize your scapulae.

Sternocleidomastoid

Pectoralis major
Serratus anterior
Trapezius
Spinal extensors
Spine
Quadratus lumborum
Iliopsoas
Rectus abdominis

Vastus medialis
Semitendinosus
Adductor magnus
Semimembranosus
Sartorius
Knee

Front **thigh**

Your **hip flexors** actively engage to stabilize you upright into hip flexion while your **hip extensors** stretch. All of your **glutes**, along with your **deep six external rotators**, stretch, especially when practising the modified variations of the pose.

» **CLOSER** LOOK

King Pigeon is challenging for some, but you can find a relaxed variation by lying down or using props. These options may relieve pressure your joints.

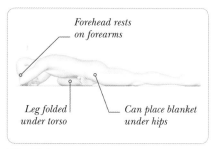

Forehead rests on forearms

Leg folded under torso

Can place blanket under hips

VARIATION

For a more passive version, release forwards. You may feel enough of a stretch on your hands or forearms. Consider using blankets or a bolster under your hips. Or, you can get similar benefits by lying on your back and placing your legs in a figure 4 position.

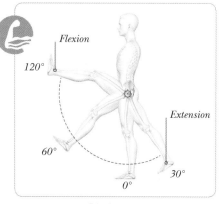

Flexion

120°

60°

0°

Extension

30°

Piriformis

Your piriformis normally externally rotates your hip. However, when your hip is flexed past 60°, your piriformis transforms action to an internal rotator. This means it stretches deeply when in external rotation and flexion, like in the front hip of many versions of Pigeon.

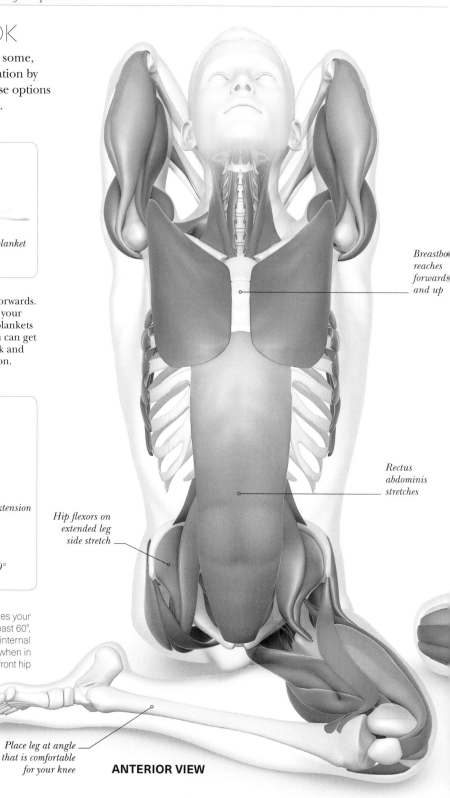

Breastbone reaches forwards and up

Rectus abdominis stretches

Hip flexors on extended leg side stretch

Place leg at angle that is comfortable for your knee

ANTERIOR VIEW

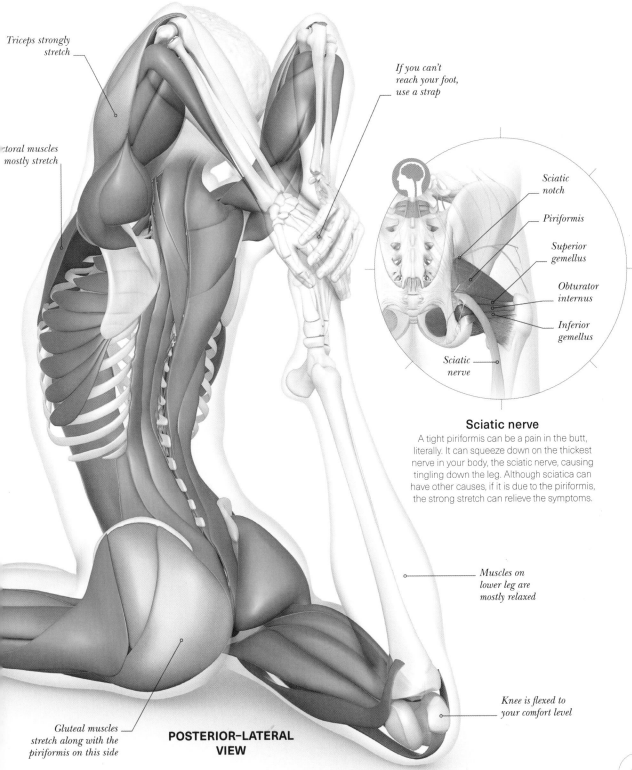

Triceps strongly stretch

If you can't reach your foot, use a strap

...toral muscles mostly stretch

Sciatic notch

Piriformis

Superior gemellus

Obturator internus

Inferior gemellus

Sciatic nerve

Sciatic nerve

A tight piriformis can be a pain in the butt, literally. It can squeeze down on the thickest nerve in your body, the sciatic nerve, causing tingling down the leg. Although sciatica can have other causes, if it is due to the piriformis, the strong stretch can relieve the symptoms.

Muscles on lower leg are mostly relaxed

Gluteal muscles stretch along with the piriformis on this side

POSTERIOR–LATERAL VIEW

Knee is flexed to your comfort level

83

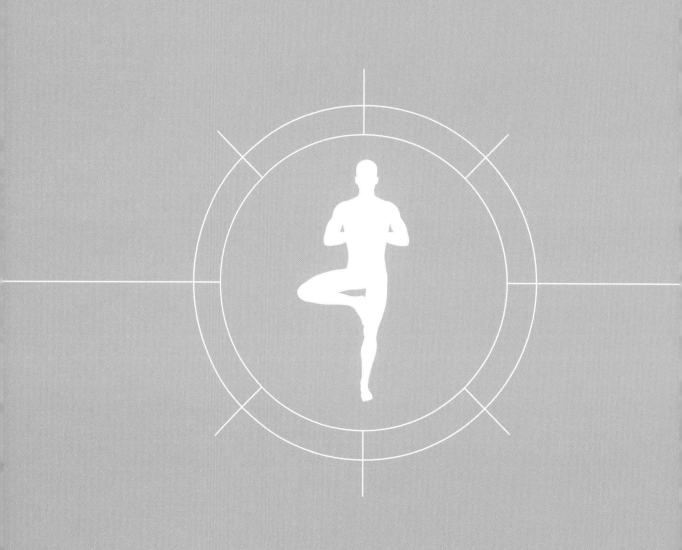

STANDING
ASANAS

These standing asanas were specifically chosen to help improve posture and balance. How you hold your body affects all the systems of your anatomy, as well as your energy levels, your cognition, and your confidence. The intention behind these poses is less pain, fewer injuries, improved posture, and optimal movement in everything you do.

MOUNTAIN
Tadasana

This standing pose is essentially the anatomical position. It represents how you hold yourself in the world – your postural alignment. The pose creates a stable connection to the earth. Many muscles are slightly engaged to support you upright, resisting gravity.

THE BIG PICTURE

Although the aim is to activate as few muscles as little as possible, a lot of muscles in your body engage subtly in a neutral or lenthening position to prevent you from leaning or falling in any direction. Your lower legs, thighs, hips, back muscles, and abdominals may all be felt buzzing with this slight engagement.

Torso

Your **spinal extensors** and **transversus abdominis** engage to lengthen and stabilize your spine. Your **rhomboids** and middle and lower **trapezius** engage to stabilize your scapulae in place. Your **pectoralis minor** may be engaging to lift your ribs.

Neck

Your **cervical extensors** engage while in a lengthened or neutral position to keep your neck long with a neutral, inward curve.

Splenius muscles

Arms

Your **posterior deltoids** slightly engage to externally rotate your shoulders, while your **anterior deltoids** stretch. Your **supinators** engage to make your palms face forward.

Shoulder

Deltoids

Elbow

Supinator

Spine

Spinal extensors

Quadratus lumborum

Transversus abdominis

Rectus abdominis

Pectoralis minor

Rhomboids

Trapezius

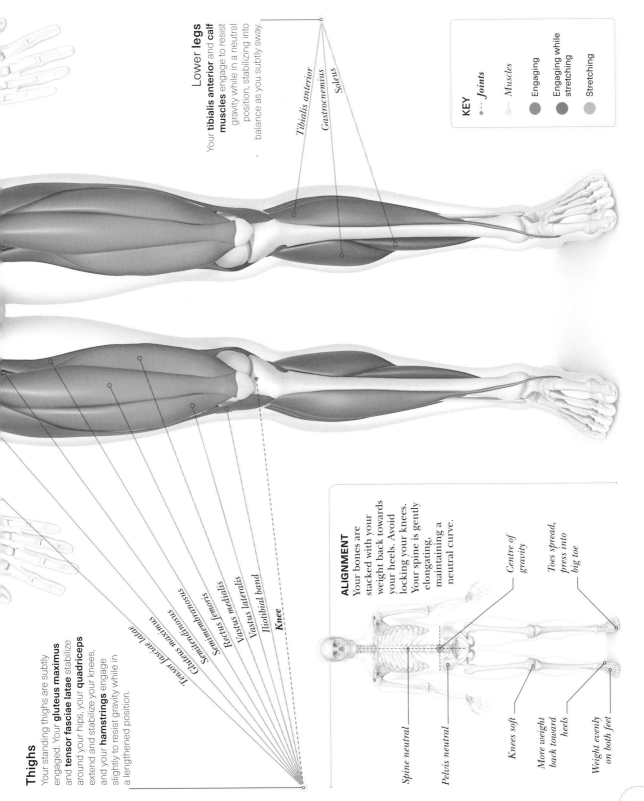

Lower legs

Your **tibialis anterior** and **calf muscles** engage to resist gravity while in a neutral position, stabilizing into balance as you subtly sway.

Tibialis anterior

Gastrocnemius

Soleus

Thighs

Your standing thighs are subtly engaged. Your **gluteus maximus** and **tensor fasciae latae** stabilize around your hips, your **quadriceps** extend and stabilize your knees, and your **hamstrings** engage slightly to resist gravity while in a lengthened position.

Tensor fasciae latae

Gluteus maximus

Semitendinosus

Semimembranosus

Rectus femoris

Vastus medialis

Vastus lateralis

Iliotibial band

Knee

ALIGNMENT

Your bones are stacked with your weight back towards your heels. Avoid locking your knees. Your spine is gently elongating, maintaining a neutral curve.

Centre of gravity

Toes spread, press into big toe

Spine neutral

Pelvis neutral

Knees soft

More weight back toward heels

Weight evenly on both feet

87

»CLOSER LOOK

Mountain pose is an opportunity to find a stable, structurally sound base. The structure and placement of your feet can facilitate the foundation of that base.

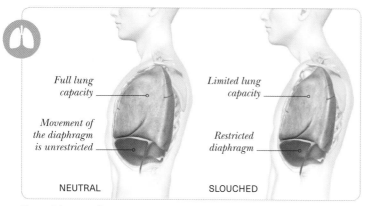

Full lung capacity

Movement of the diaphragm is unrestricted

NEUTRAL

Limited lung capacity

Restricted diaphragm

SLOUCHED

Breathing and posture

When you slouch, you have limited lung capacity, along with restricted movement of your diaphragm. From a yoga perspective, when you aren't breathing well, your prana, or vital energy, is not flowing properly. From a physiological perspective, when your respiratory system is not efficient, neither are your cardiovascular, digestive, endocrine, or nervous systems. So, stand up tall and let your body function optimally.

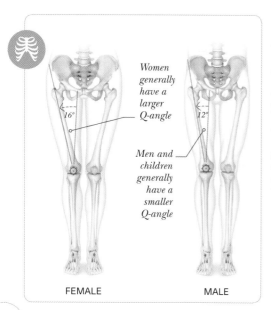

Women generally have a larger Q-angle

16°

Men and children generally have a smaller Q-angle

12°

FEMALE

MALE

Feet at hip distance

Some styles of yoga bring the feet together in Tadasana. However, while many modern asanas were developed for pre-adolescent boys in India, who have fairly narrow hips, yoga is now predominantly practised by adult women, whose hips are wider. For many people, standing with the feet at hip distance is more stable, decreasing the Q-angle (shown left) and reducing stress on the knees.

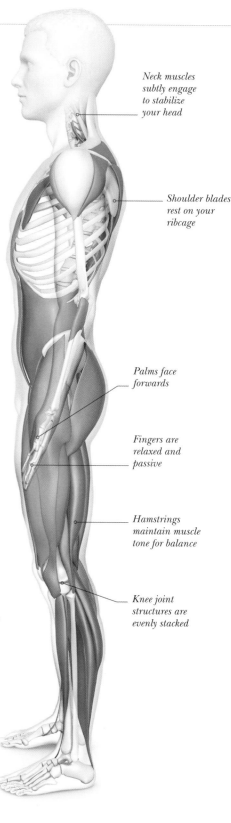

Neck muscles subtly engage to stabilize your head

Shoulder blades rest on your ribcage

Palms face forwards

Fingers are relaxed and passive

Hamstrings maintain muscle tone for balance

Knee joint structures are evenly stacked

LATERAL VIEW

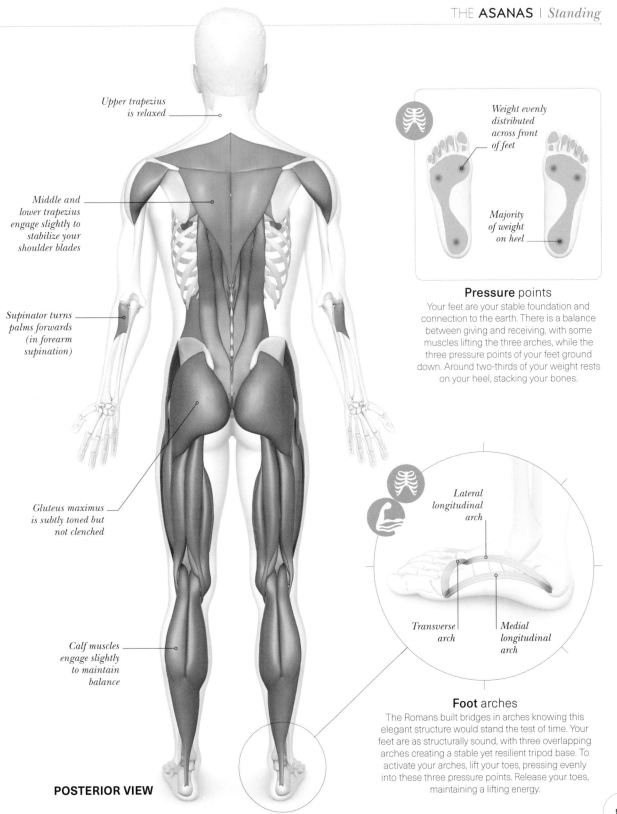

*Upper trapezius
is relaxed*

*Middle and
lower trapezius
engage slightly to
stabilize your
shoulder blades*

*Supinator turns
palms forwards
(in forearm
supination)*

*Gluteus maximus
is subtly toned but
not clenched*

*Calf muscles
engage slightly
to maintain
balance*

POSTERIOR VIEW

*Weight evenly
distributed
across front
of feet*

*Majority
of weight
on heel*

Pressure points
Your feet are your stable foundation and
connection to the earth. There is a balance
between giving and receiving, with some
muscles lifting the three arches, while the
three pressure points of your feet ground
down. Around two-thirds of your weight rests
on your heel, stacking your bones.

*Lateral
longitudinal
arch*

*Transverse
arch*

*Medial
longitudinal
arch*

Foot arches
The Romans built bridges in arches knowing this
elegant structure would stand the test of time. Your
feet are as structurally sound, with three overlapping
arches creating a stable yet resilient tripod base. To
activate your arches, lift your toes, pressing evenly
into these three pressure points. Release your toes,
maintaining a lifting energy.

FORWARD FOLD

Uttanasana

Forward fold offers an opportunity to improve flexibility. Transitioning in and out of the pose, such as in sun salutations, will help prepare you for common functional movements you do throughout the day. This pose can be adapted for all abilities by going into the fold less deeply.

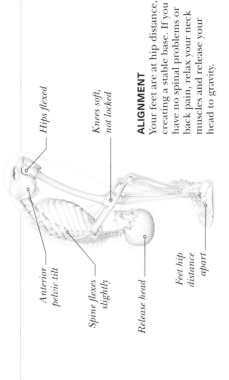

Hips flexed

Knees soft, not locked

ALIGNMENT
Your feet are at hip distance, creating a stable base. If you have no spinal problems or back pain, relax your neck muscles and release your head to gravity.

Anterior pelvic tilt

Spine flexes slightly

Release head

Feet hip distance apart

THE BIG PICTURE

The back of your whole body is stretching – including your calf muscles, thighs, buttocks, and back muscles. At the front of your body – especially in your legs – your muscles are working to stabilize you in the deep bend.

Neck and **torso**
All of your **spinal extensors** and your **latissimus dorsi** stretch when you release your upper body to gravity.

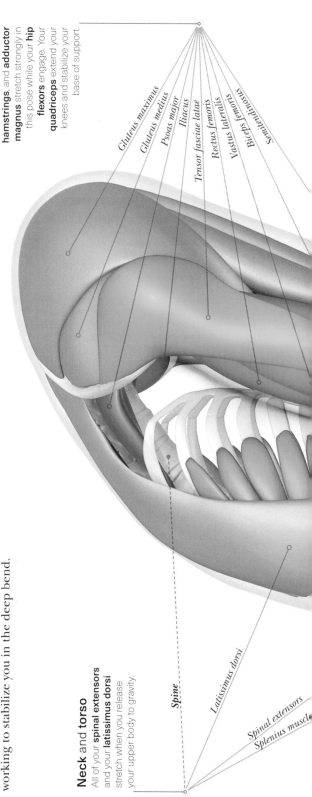

Thighs
Your **gluteus maximus, medius,** and **minimus, hamstrings,** and **adductor magnus** stretch strongly in this pose while your **hip flexors** engage. Your **quadriceps** extend your knees and stabilize your base of support.

Gluteus maximus

Gluteus medius

Psoas major

Iliacus

Tensor fasciae latae

Rectus femoris

Vastus lateralis

Biceps femoris

Semitendinosus

Spine

Latissimus dorsi

Spinal extensors

Splenius muscle

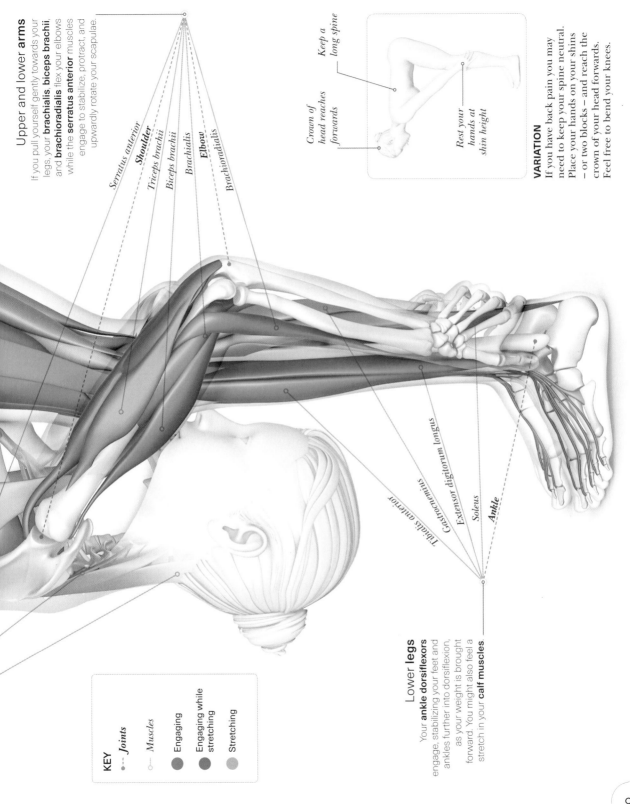

Upper and lower arms

If you pull yourself gently towards your legs, your **brachialis**, **biceps brachii**, and **brachioradialis** flex your elbows while the **serratus anterior** muscles engage to stabilize, protract, and upwardly rotate your scapulae.

Serratus anterior

Shoulder

Triceps brachii

Biceps brachii

Brachialis

Elbow

Brachioradialis

Keep a
long spine

Crown of
head reaches
forwards

Rest your
hands at
shin height

VARIATION

If you have back pain you may need to keep your spine neutral. Place your hands on your shins – or two blocks – and reach the crown of your head forwards. Feel free to bend your knees.

Tibialis anterior

Gastrocnemius

Extensor digitorum longus

Soleus

Ankle

Lower legs

Your **ankle dorsiflexors** engage, stabilizing your feet and ankles further into dorsiflexion, as your weight is brought forward. You might also feel a stretch in your **calf muscles**.

KEY

- - - Joints
- ○— Muscles

● Engaging

● Engaging while stretching

● Stretching

»» CLOSER LOOK

Forward Fold delivers a deep spinal stretch, which can help to improve back health and reduce back pain. However, care should be taken to reduce the lumbar load for those with intervertebral disc issues.

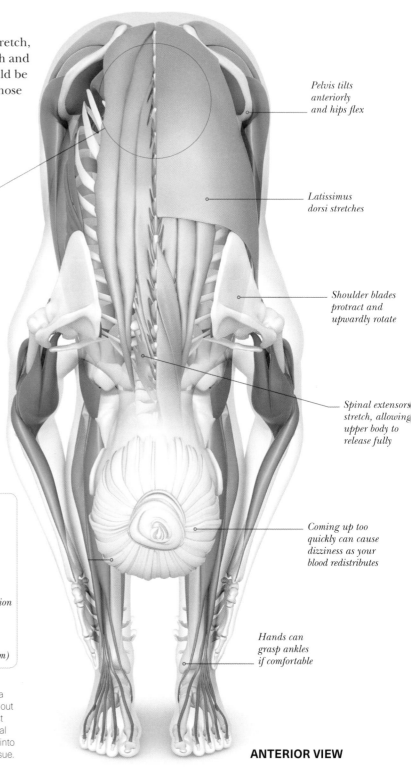

Pelvis tilts anteriorly and hips flex

Latissimus dorsi stretches

Shoulder blades protract and upwardly rotate

Spinal extensors stretch, allowing upper body to release fully

Coming up too quickly can cause dizziness as your blood redistributes

Hands can grasp ankles if comfortable

ANTERIOR VIEW

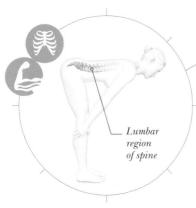

Lumbar region of spine

Lumbar load

The load on the lumbar spine in a standing Forward Fold is significant. The lower back is particularly vulnerable during the transition in and out of the pose. If you have any back pain, arthritis, disc issues, osteopenia, or osteoporosis, try keeping your spine neutral and transition in and out of the pose with bent knees and an engaged core.

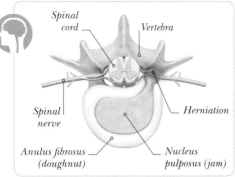

Spinal cord

Vertebra

Herniation

Spinal nerve

Anulus fibrosus (doughnut)

Nucleus pulposus (jam)

Herniated disc

Intervertebral discs are like jam doughnuts. In a "slipped" or herniated disc, the "jam" partially leaks out of the tougher fibrocartilage "dough". Since most herniations occur posterior-laterally due to spinal flexion, move slowly or avoid flexion by not going into the pose as deeply if you currently have a disc issue.

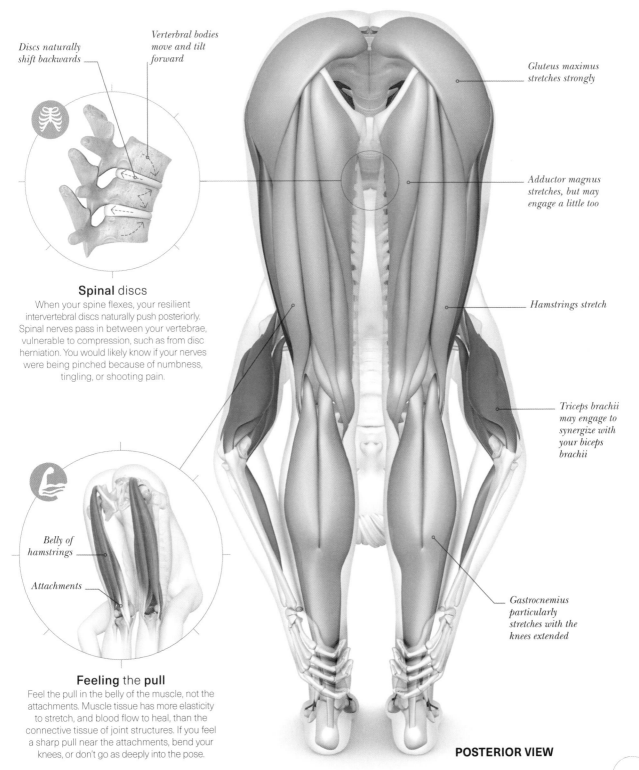

Discs naturally shift backwards

Verterbral bodies move and tilt forward

Gluteus maximus stretches strongly

Adductor magnus stretches, but may engage a little too

Hamstrings stretch

Triceps brachii may engage to synergize with your biceps brachii

Gastrocnemius particularly stretches with the knees extended

Spinal discs

When your spine flexes, your resilient intervertebral discs naturally push posteriorly. Spinal nerves pass in between your vertebrae, vulnerable to compression, such as from disc herniation. You would likely know if your nerves were being pinched because of numbness, tingling, or shooting pain.

Belly of hamstrings

Attachments

Feeling the pull

Feel the pull in the belly of the muscle, not the attachments. Muscle tissue has more elasticity to stretch, and blood flow to heal, than the connective tissue of joint structures. If you feel a sharp pull near the attachments, bend your knees, or don't go as deeply into the pose.

POSTERIOR VIEW

93

CHAIR
Utkatasana

Chair pose activates the largest muscles in your body, gets your heart pumping, and engages your core strongly. This energizing standing pose improves your thigh strength, which some studies suggest is a key factor in prolonging your life.

THE BIG PICTURE

Muscles around your thighs, hips, and core are engaging strongly to hold you in this squatting position. Lifting your arms overhead further challenges your core strength and engages your shoulder muscles. Alternatively, you can put your hands on your hips to lighten the load.

Neck

Although your **upper trapezius** engages slightly to elevate your scapulae, aim to consciously soften the area, letting go of extraneous tension. Your **cervical extensors** engage to prevent your head from dropping forwards.

Arms

Your **shoulder flexors** engage to bring your arms overhead. Your **deltoids** dynamically engage to abduct your arms into position, and to help hold your arms in shoulder flexion. Your **triceps** extend your elbows.

Torso

Your **spinal extensors** and **transversus abdominis** engage to stabilize your spine in neutral curves. Your **rectus abdominis** is mostly lengthening. Your **rhomboids** engage with your **middle** and **lower trapezius** to retract and stabilize your scapulae. Your **latissimus dorsi** stretches with shoulder flexion.

Brachioradialis

Elbow

Brachialis

Triceps brachii

Shoulder

Deltoids

Pectoralis major

Serratus anterior

Cervical extensors

Trapezius (mid/lower)

Latissimus dorsi

Spine

Quadratus lumborum

Rectus abdominis

Transversus ab

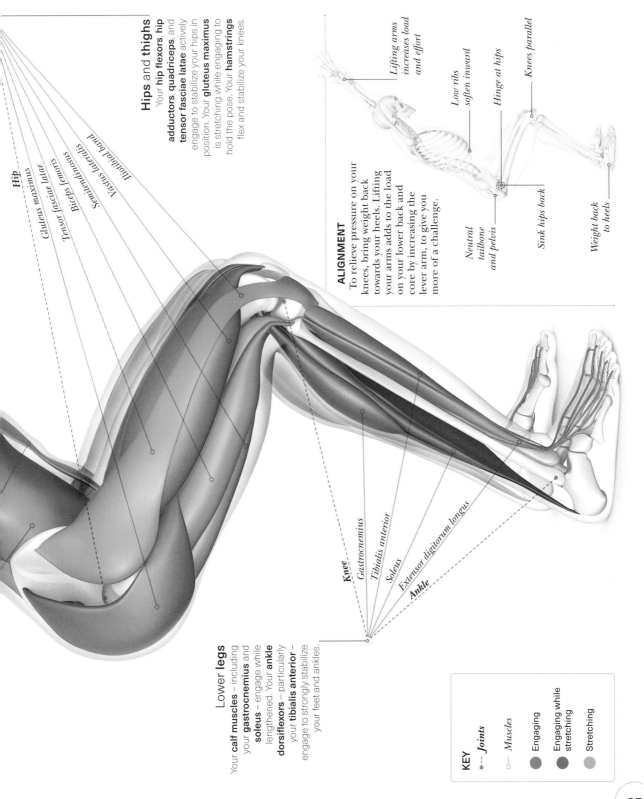

Hips and thighs

Your **hip flexors**, **hip adductors**, **quadriceps**, and **tensor fasciae latae** actively engage to stabilize your hips in position. Your **gluteus maximus** is stretching while engaging to hold the pose. Your **hamstrings** flex and stabilize your knees.

Hip

Gluteus maximus
Tensor fasciae latae
Biceps femoris
Semitendinosus
Vastus lateralis
Iliotibial band

ALIGNMENT

To relieve pressure on your knees, bring weight back towards your heels. Lifting your arms adds to the load on your lower back and core by increasing the lever arm, to give you more of a challenge.

Lifting arms increases load and effort

Low ribs soften inward

Knees parallel

Hinge at hips

Neutral tailbone and pelvis

Sink hips back

Weight back to heels

Lower legs

Your **calf muscles** – including your **gastrocnemius** and **soleus** – engage while lengthened. Your **ankle dorsiflexors** – particularly your **tibialis anterior** – engage to strongly stabilize your feet and ankles.

Knee
Gastrocnemius
Tibialis anterior
Soleus
Extensor digitorum longus
Ankle

KEY

- - - *Joints*
— ○ — *Muscles*

● Engaging
● Engaging while stretching
● Stretching

❯❯ CLOSER LOOK

Chair pose leads to body-wide effects.
Lifting your arms, for example, raises
blood pressure. Raising your arms also
increases the lumbar load, which tests
your cardiovascular system and core muscles.

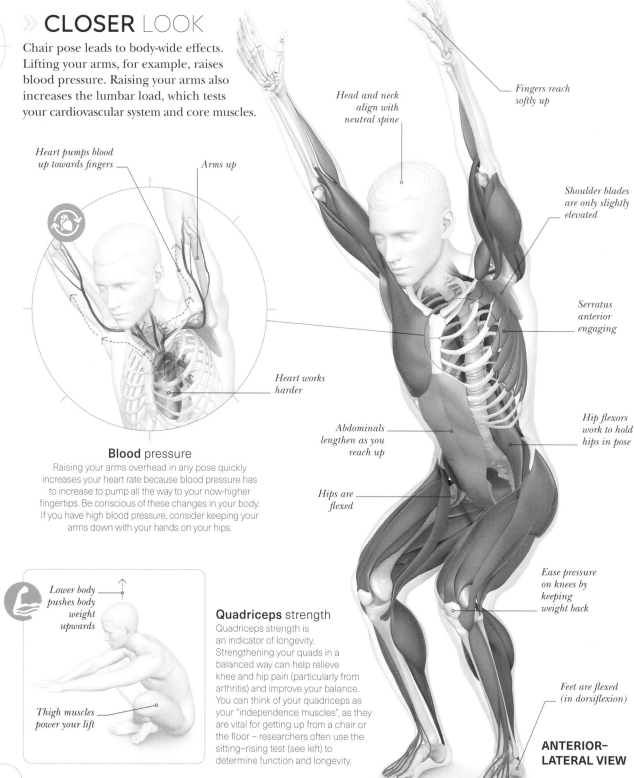

*Heart pumps blood
up towards fingers*

Arms up

*Head and neck
align with
neutral spine*

*Fingers reach
softly up*

*Shoulder blades
are only slightly
elevated*

*Serratus
anterior
engaging*

*Heart works
harder*

*Abdominals
lengthen as you
reach up*

*Hip flexors
work to hold
hips in pose*

*Hips are
flexed*

Blood pressure

Raising your arms overhead in any pose quickly
increases your heart rate because blood pressure has
to increase to pump all the way to your now-higher
fingertips. Be conscious of these changes in your body.
If you have high blood pressure, consider keeping your
arms down with your hands on your hips.

*Ease pressure
on knees by
keeping
weight back*

*Lower body
pushes body
weight
upwards*

Quadriceps strength

Quadriceps strength is
an indicator of longevity.
Strengthening your quads in a
balanced way can help relieve
knee and hip pain (particularly from
arthritis) and improve your balance.
You can think of your quadriceps as
your "independence muscles", as they
are vital for getting up from a chair or
the floor – researchers often use the
sitting–rising test (see left) to
determine function and longevity.

*Thigh muscles
power your lift*

*Feet are flexed
(in dorsiflexion)*

**ANTERIOR–
LATERAL VIEW**

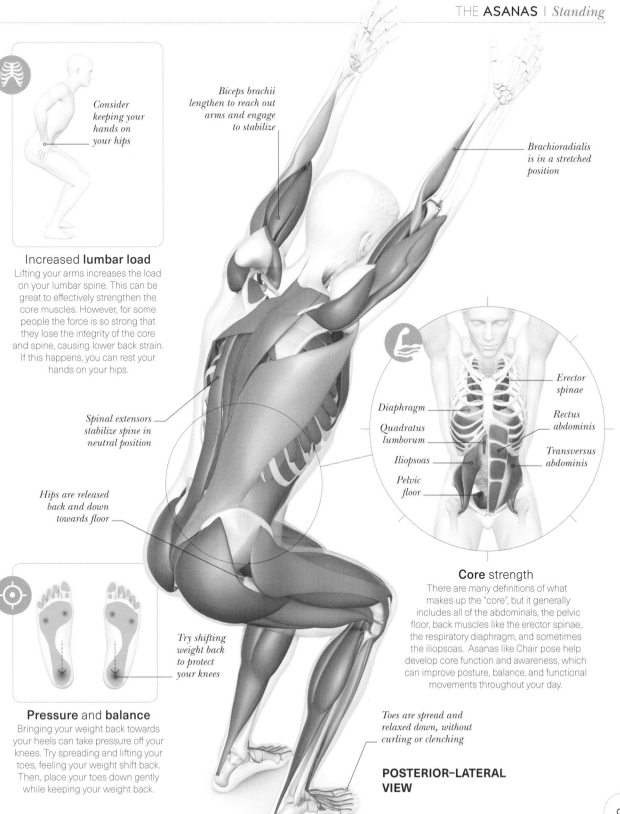

Consider keeping your hands on your hips

Biceps brachii lengthen to reach out arms and engage to stabilize

Brachioradialis is in a stretched position

Increased **lumbar load**

Lifting your arms increases the load on your lumbar spine. This can be great to effectively strengthen the core muscles. However, for some people the force is so strong that they lose the integrity of the core and spine, causing lower back strain. If this happens, you can rest your hands on your hips.

Spinal extensors stabilize spine in neutral position

Hips are released back and down towards floor

Diaphragm

Quadratus lumborum

Iliopsas

Pelvic floor

Erector spinae

Rectus abdominis

Transversus abdominis

Core strength

There are many definitions of what makes up the "core", but it generally includes all of the abdominals, the pelvic floor, back muscles like the erector spinae, the respiratory diaphragm, and sometimes the iliopsoas. Asanas like Chair pose help develop core function and awareness, which can improve posture, balance, and functional movements throughout your day.

Try shifting weight back to protect your knees

Pressure and **balance**

Bringing your weight back towards your heels can take pressure off your knees. Try spreading and lifting your toes, feeling your weight shift back. Then, place your toes down gently while keeping your weight back.

Toes are spread and relaxed down, without curling or clenching

POSTERIOR–LATERAL VIEW

97

CRESCENT LUNGE

Anjaneyasana

This lunge is a good antidote to sitting down too much. It is also beneficial for runners or anyone who participates in sports that involve running, because it strengthens the muscles that power your stride and stretches your hip flexors.

THE BIG PICTURE

In this pose, the muscles of your hips and your gluteus muscles stretch and activate dynamically to keep you balanced. Your thigh muscles engage strongly to stabilize your hips and knees, while your core muscles stabilize your spine in a slight backbend.

Neck

Your **cervical extensors** engage to extend your cervical spine while your **cervical flexors** engage and lengthen to stabilize your neck, preventing your head from dropping back.

Scissor thighs together

Slight extension of spine

Lift chin slightly

Knee directly over ankle

Press down into feet to feel core engage

Feet hip distance apart

On ball of foot

ALIGNMENT
Your feet are hip width apart to maintain balance. Your front knee is directly over your ankle or behind it for safety.

Elbow

Brachioradialis

Brachialis

Biceps brachii

Coracobrachialis

Deltoids

Shoulder

Serratus anterior

Arms

Your **shoulder flexors** engage. Your **anterior deltoids** aid in shoulder flexion while your **posterior deltoids** lengthen, yet some fibres engage to stabilize and externally rotate your shoulders. Your **triceps** extend your elbows. Feel space, not stiffness, in your joints as you reach through your fingers.

Longus muscles

Sternocleidomastoid

Splenius muscles

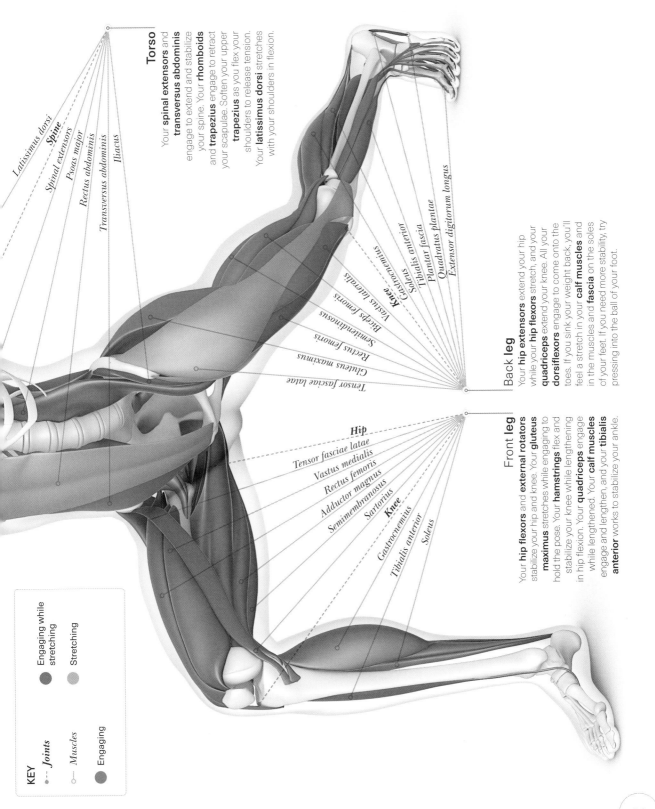

Torso

Your **spinal extensors** and **transversus abdominis** engage to extend and stabilize your spine. Your **rhomboids** and **trapezius** engage to retract your scapulae. Soften your upper **trapezius** as you flex your shoulders to release tension. Your **latissimus dorsi** stretches with your shoulders in flexion.

Spine

Latissimus dorsi
Spinal extensors
Psoas major
Rectus abdominis
Transversus abdominis
Iliacus

Back leg

Your **hip extensors** extend your hip while your **hip flexors** stretch, and your **quadriceps** extend your knee. All your **dorsiflexors** engage to come onto the toes. If you sink your weight back, you'll feel a stretch in your **calf muscles** and **fascia** on the soles of your feet. If you need more stability, try pressing into the ball of your foot.

Knee

Gastrocnemius
Soleus anterior
Tibialis fascia
Plantar fascia
Quadratus plantae
Extensor digitorum longus

Vastus lateralis
Biceps femoris
Semitendinosus
Rectus femoris
Gluteus maximus
Tensor fasciae latae

Front leg

Your **hip flexors** and **external rotators** stabilize your hip and knee. Your **gluteus maximus** stretches while engaging to hold the pose. Your **hamstrings** flex and stabilize your knee while lengthening in hip flexion. Your **quadriceps** engage while lengthened. Your **calf muscles** engage and lengthen, and your **tibialis anterior** works to stabilize your ankle.

Hip

Tensor fasciae latae
Vastus medialis
Rectus femoris
Adductor magnus
Semimembranosus
Sartorius

Knee

Gastrocnemius
Tibialis anterior
Soleus

KEY

Joints

Muscles

Engaging

Engaging while stretching

Stretching

CRESCENT LUNGE | Anjaneyasana

» CLOSER LOOK

You may try modifications to find comfort and efficient alignment. This pose presents an opportunity to consciously release common "stress" and "fear" muscles like the upper trapezius and psoas major.

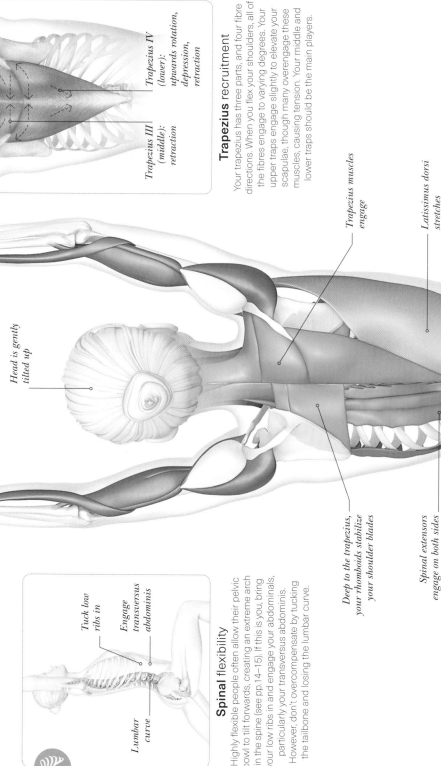

Trapezius I (upper): shoulder blade elevation

Trapezius II (middle): elevation, upwards rotation, retraction

Trapezius III (middle): retraction

Trapezius IV (lower): upwards rotation, depression, retraction

Trapezius recruitment

Your trapezius has three parts, and four fibre directions. When you flex your shoulders, all of the fibres engage to varying degrees. Your upper traps engage slightly to elevate your scapulae, though many overengage these muscles, causing tension. Your middle and lower traps should be the main players.

Trapezius muscles engage

Latissimus dorsi stretches

Fingers softly reach up

Head is gently tilted up

Deep to the trapezius, your rhomboids stabilize your shoulder blades

Spinal extensors engage on both sides

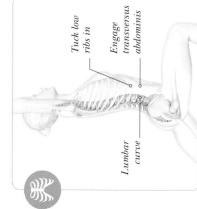

Tuck low ribs in

Engage transversus abdominis

Lumbar curve

Spinal flexibility

Highly flexible people often allow their pelvic bowl to tilt forwards, creating an extreme arch in the spine (see pp.14–15). If this is you, bring your low ribs in and engage your abdominals, particularly your transversus abdominis. However, don't overcompensate by tucking the tailbone and losing the lumbar curve.

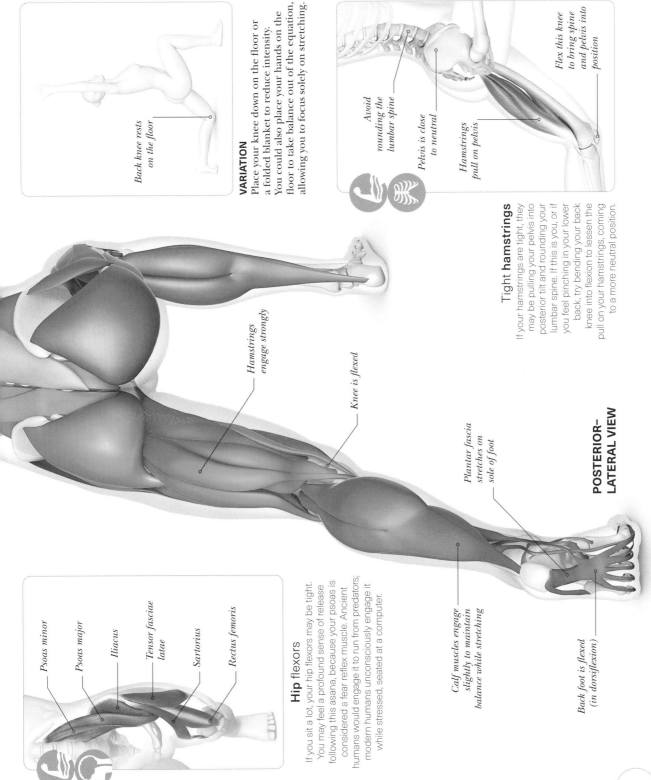

VARIATION

Place your knee down on the floor or a folded blanket to reduce intensity. You could also place your hands on the floor to take balance out of the equation, allowing you to focus solely on stretching.

Back knee rests on the floor

Flex this knee to bring spine and pelvis into position

Avoid rounding the lumbar spine

Pelvis is close to neutral

Hamstrings pull on pelvis

Tight hamstrings

If your hamstrings are tight, they may be pulling your pelvis into posterior tilt and rounding your lumbar spine. If this is you, or if you feel pinching in your lower back, try bending your back knee into flexion to lessen the pull on your hamstrings, coming to a more neutral position.

Hamstrings engage strongly

Knee is flexed

Plantar fascia stretches on sole of foot

POSTERIOR–LATERAL VIEW

Calf muscles engage slightly to maintain balance while stretching

Back foot is flexed (in dorsiflexion)

Hip flexors

If you sit a lot, your hip flexors may be tight. You may feel a profound sense of release following this asana, because your psoas is considered a fear reflex muscle. Ancient humans would engage it to run from predators; modern humans unconsciously engage it while stressed, seated at a computer.

Psoas minor

Psoas major

Iliacus

Tensor fasciae latae

Sartorius

Rectus femoris

WARRIOR II
Virabhadrasana II

This strong standing pose is grounding, energizing, and stabilizing. Holding Warrior II for a period of time works on your balance and muscular strength, and in doing so provides a great opportunity to observe how your mind reacts during a heated challenge.

THE BIG PICTURE

This pose engages large muscles around your thighs and core. Your arms are reaching in both directions, creating space in the joints, without stiffening or locking your elbows or fingers.

KEY

- **- *Joints***

o— *Muscles*

● Engaging

● Engaging while stretching

● Stretching

Wrist

Brachioradialis

Pronator quadratus

Elbow

Biceps brachii

Deltoids

Rotator cuff muscles

Pectoralis minor

Serratus anterior

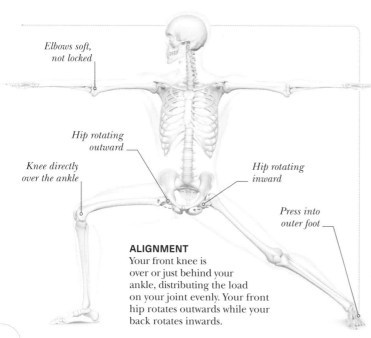

Elbows soft, not locked

Hip rotating outward

Knee directly over the ankle

Hip rotating inward

Press into outer foot

ALIGNMENT
Your front knee is over or just behind your ankle, distributing the load on your joint evenly. Your front hip rotates outwards while your back rotates inwards.

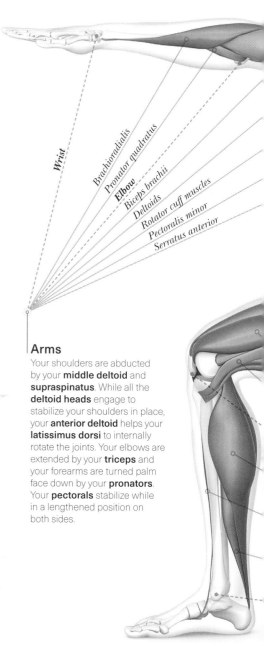

Arms
Your shoulders are abducted by your **middle deltoid** and **supraspinatus**. While all the **deltoid heads** engage to stabilize your shoulders in place, your **anterior deltoid** helps your **latissimus dorsi** to internally rotate the joints. Your elbows are extended by your **triceps** and your forearms are turned palm face down by your **pronators**. Your **pectorals** stabilize while in a lengthened position on both sides.

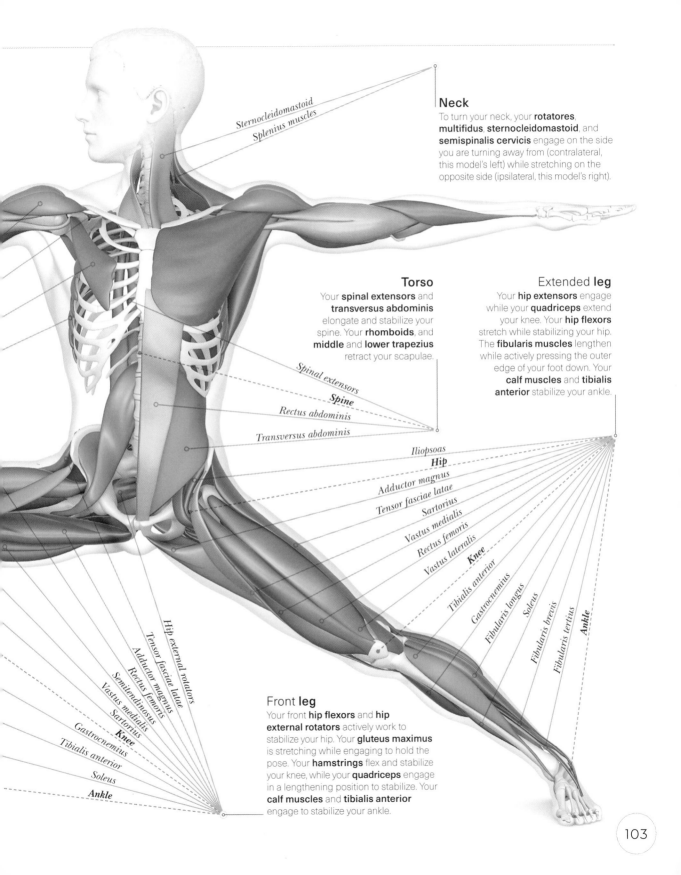

Neck

To turn your neck, your **rotatores**, **multifidus**, **sternocleidomastoid**, and **semispinalis cervicis** engage on the side you are turning away from (contralateral, this model's left) while stretching on the opposite side (ipsilateral, this model's right).

Sternocleidomastoid

Splenius muscles

Torso

Your **spinal extensors** and **transversus abdominis** elongate and stabilize your spine. Your **rhomboids**, and **middle** and **lower trapezius** retract your scapulae.

Spinal extensors

Spine

Rectus abdominis

Transversus abdominis

Extended **leg**

Your **hip extensors** engage while your **quadriceps** extend your knee. Your **hip flexors** stretch while stabilizing your hip. The **fibularis muscles** lengthen while actively pressing the outer edge of your foot down. Your **calf muscles** and **tibialis anterior** stabilize your ankle.

Iliopsoas

Hip

Adductor magnus

Tensor fasciae latae

Sartorius

Vastus medialis

Rectus femoris

Vastus lateralis

Knee

Tibialis anterior

Gastrocnemius

Fibularis longus

Soleus

Fibularis brevis

Fibularis tertius

Ankle

Hip external rotators

Tensor fasciae latae

Adductor magnus

Rectus femoris

Semitendinosus

Vastus medialis

Sartorius

Knee

Gastrocnemius

Tibialis anterior

Soleus

Ankle

Front **leg**

Your front **hip flexors** and **hip external rotators** actively work to stabilize your hip. Your **gluteus maximus** is stretching while engaging to hold the pose. Your **hamstrings** flex and stabilize your knee, while your **quadriceps** engage in a lengthening position to stabilize. Your **calf muscles** and **tibialis anterior** engage to stabilize your ankle.

103

WARRIOR II | *Virabhadrasana II*

» CLOSER LOOK

Proper alignment in Warrior II can prevent damage to joint structures, especially in your knees. This is vital because the knee is one of the most mechanically complex joints in the body.

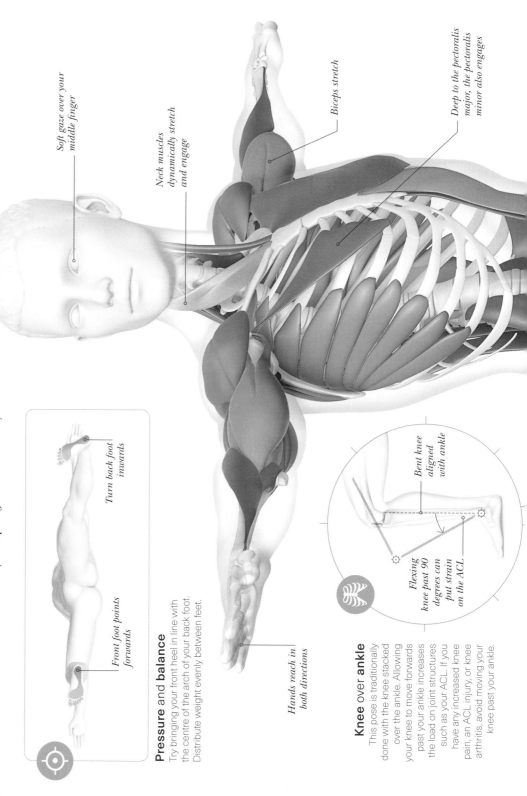

Soft gaze over your middle finger

Neck muscles dynamically stretch and engage

Biceps stretch

Deep to the pectoralis major, the pectoralis minor also engages

Hands reach in both directions

Turn back foot inwards

Front foot points forwards

Pressure and balance

Try bringing your front heel in line with the centre of the arch of your back foot. Distribute weight evenly between feet.

Bent knee aligned with ankle

Flexing knee past 90 degrees can put strain on the ACL

Knee over ankle

This pose is traditionally done with the knee stacked over the ankle. Allowing your knee to move forwards past your ankle increases the load on joint structures such as your ACL. If you have any increased knee pain, an ACL injury, or knee arthritis, avoid moving your knee past your ankle.

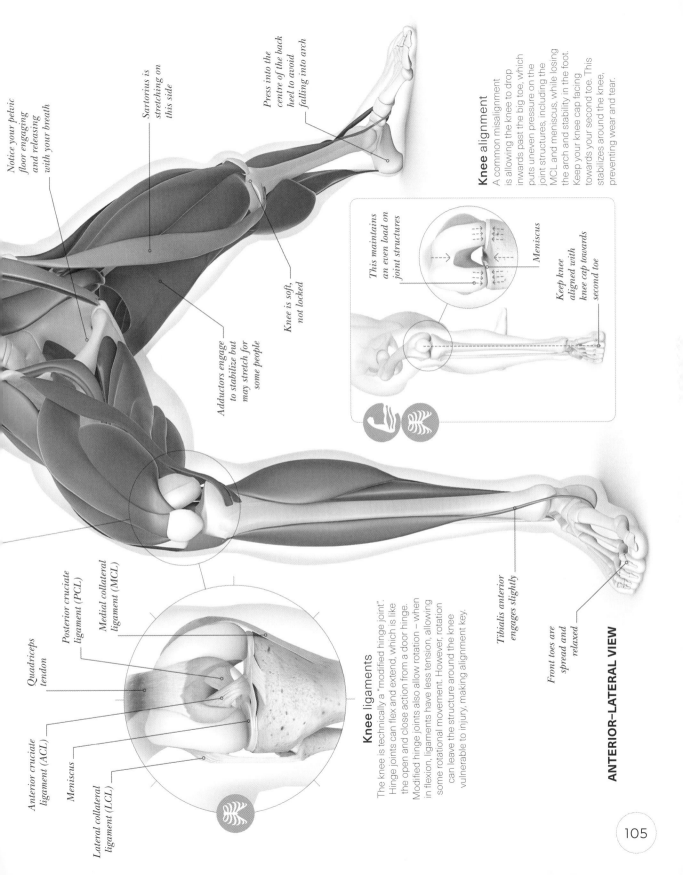

Notice your pelvic floor engaging and releasing with your breath

Sartorius is stretching on this side

Press into the centre of the back heel to avoid falling into arch

Adductors engage to stabilize but may stretch for some people

Knee is soft, not locked

Knee alignment

A common misalignment is allowing the knee to drop inwards past the big toe, which puts uneven pressure on the joint structures, including the MCL and meniscus, while losing the arch and stability in the foot. Keep your knee cap facing towards your second toe. This stabilizes around the knee, preventing wear and tear.

This maintains an even load on joint structures

Meniscus

Keep knee aligned with knee cap towards second toe

Quadriceps tendon

Posterior cruciate ligament (PCL)

Medial collateral ligament (MCL)

Anterior cruciate ligament (ACL)

Meniscus

Lateral collateral ligament (LCL)

Tibialis anterior engages slightly

Front toes are spread and relaxed

Knee ligaments

The knee is technically a "modified hinge joint". Hinge joints can flex and extend, which is like the open and close action from a door hinge. Modified hinge joints also allow rotation – when in flexion, ligaments have less tension, allowing some rotational movement. However, rotation can leave the structure around the knee vulnerable to injury, making alignment key.

ANTERIOR–LATERAL VIEW

WARRIOR III
Virabhadrasana III

Warrior III is a strong, standing balancing pose that increases your focus and co-ordination. Your balance is particularly challenged as you bring your head parallel to the ground, affecting structures inside your inner ear that monitor your position and help to keep you upright.

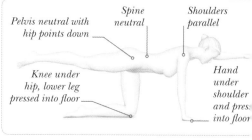

Labels: *Pelvis neutral with hip points down* — *Spine neutral* — *Shoulders parallel* — *Knee under hip, lower leg pressed into floor* — *Hand under shoulder and pres. into floor*

VARIATION
Sunbird challenges your balance but from a more stable base. Start on all fours, then lift an arm at shoulder height and the opposite leg at hip height.

THE BIG PICTURE

The muscles of your thighs, lower legs, and ankles strengthen as you try to maintain your balance on one leg. Muscles around your hips, core, and shoulders work hard to hold the rest of your body horizontal.

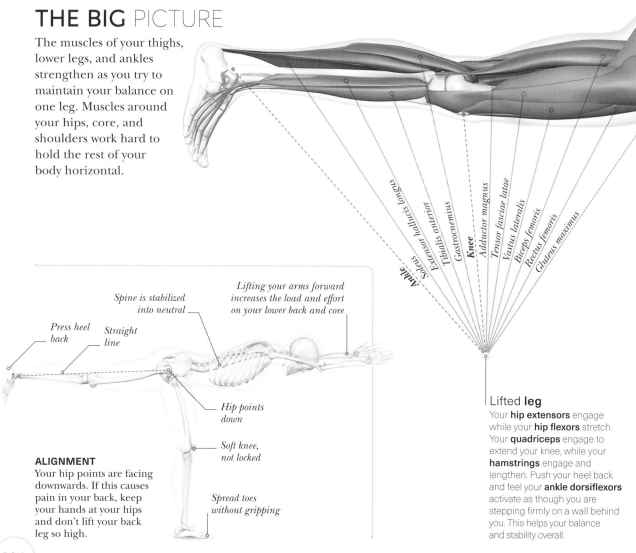

Muscle labels: **Ankle** · Soleus · Extensor hallucis longus · Tibialis anterior · Gastrocnemius · **Knee** · Adductor magnus · Tensor fasciae latae · Vastus lateralis · Biceps femoris · Rectus femoris · Gluteus maximus

Labels: *Press heel back* — *Straight line* — *Spine is stabilized into neutral* — *Lifting your arms forward increases the load and effort on your lower back and core* — *Hip points down* — *Soft knee, not locked* — *Spread toes without gripping*

ALIGNMENT
Your hip points are facing downwards. If this causes pain in your back, keep your hands at your hips and don't lift your back leg so high.

Lifted leg
Your **hip extensors** engage while your **hip flexors** stretch. Your **quadriceps** engage to extend your knee, while your **hamstrings** engage and lengthen. Push your heel back and feel your **ankle dorsiflexors** activate as though you are stepping firmly on a wall behind you. This helps your balance and stability overall.

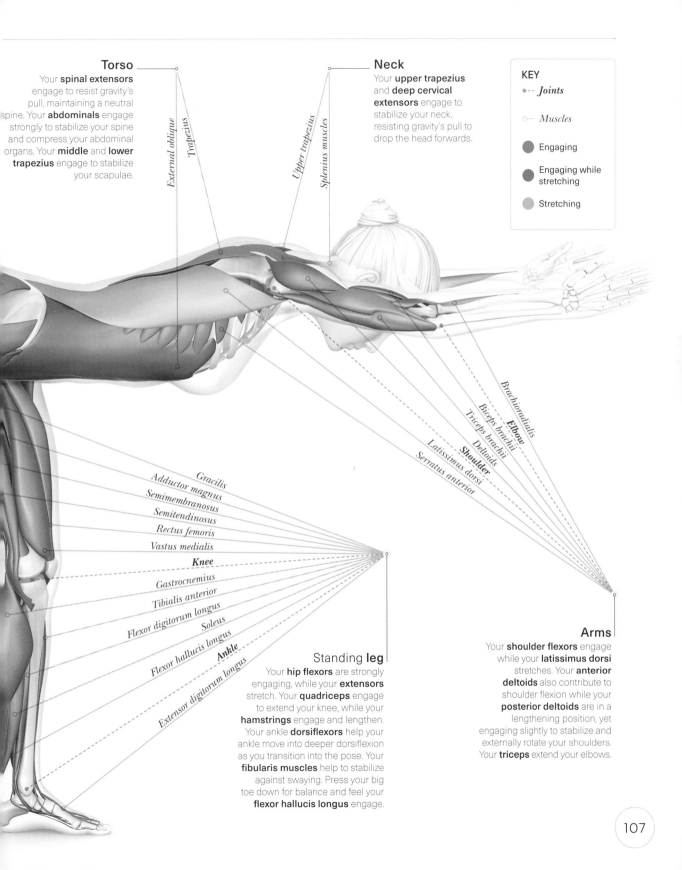

Torso

Your **spinal extensors** engage to resist gravity's pull, maintaining a neutral spine. Your **abdominals** engage strongly to stabilize your spine and compress your abdominal organs. Your **middle** and **lower trapezius** engage to stabilize your scapulae.

External oblique
Trapezius

Neck

Your **upper trapezius** and **deep cervical extensors** engage to stabilize your neck, resisting gravity's pull to drop the head forwards.

Upper trapezius
Splenius muscles

KEY

- •-- *Joints*
- ⊶ *Muscles*
- Engaging
- Engaging while stretching
- Stretching

Brachioradialis
Elbow
Biceps brachii
Triceps brachii
Deltoids
Shoulder
Latissimus dorsi
Serratus anterior

Gracilis
Adductor magnus
Semimembranosus
Semitendinosus
Rectus femoris
Vastus medialis
Knee

Gastrocnemius
Tibialis anterior
Flexor digitorum longus
Soleus
Flexor hallucis longus
Ankle
Extensor digitorum longus

Standing leg

Your **hip flexors** are strongly engaging, while your **extensors** stretch. Your **quadriceps** engage to extend your knee, while your **hamstrings** engage and lengthen. Your ankle **dorsiflexors** help your ankle move into deeper dorsiflexion as you transition into the pose. Your **fibularis muscles** help to stabilize against swaying. Press your big toe down for balance and feel your **flexor hallucis longus** engage.

Arms

Your **shoulder flexors** engage while your **latissimus dorsi** stretches. Your **anterior deltoids** also contribute to shoulder flexion while your **posterior deltoids** are in a lengthening position, yet engaging slightly to stabilize and externally rotate your shoulders. Your **triceps** extend your elbows.

»CLOSER LOOK

There are three mechanisms of balance: inner ear, visual, and proprioceptive input. Warrior III challenges each of these systems, improving your dynamic balance as you enter the pose and static balance while you hold it.

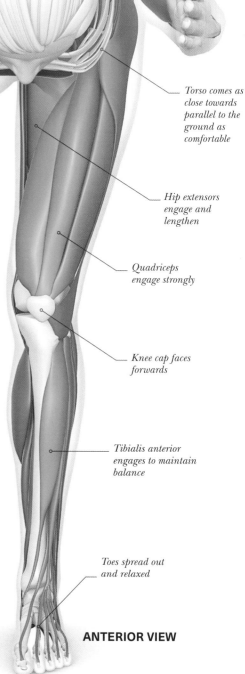

Torso comes as close towards parallel to the ground as comfortable

Hip extensors engage and lengthen

Quadriceps engage strongly

Knee cap faces forwards

Tibialis anterior engages to maintain balance

Toes spread out and relaxed

ANTERIOR VIEW

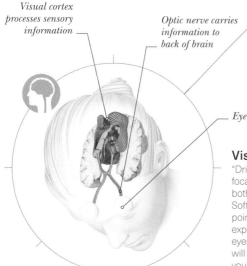

Visual cortex processes sensory information

Optic nerve carries information to back of brain

Eye

Visual input

"Drishti" is the yoga term for a focal point, which can help with both balance and concentration. Softly focus on a single stationary point ahead of you. You can also experiment with closing your eyes for a few moments – you will quickly realize how much your visual input contributes to your balance.

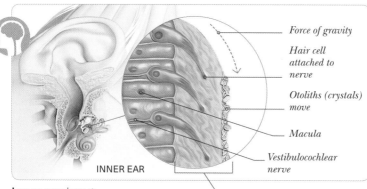

Force of gravity

Hair cell attached to nerve

Otoliths (crystals) move

Macula

Vestibulocochlear nerve

INNER EAR

Otolithic membrane contains gelatinous fluid

Inner ear input

Your inner ear has tunnels in a bony labyrinth filled with fluid to regulate your equilibrium or balance. When your head changes orientation, the fluid pushes on sensitive hair cells. Attached nerves tell your brain which direction your head is moving, to adjust for balance.

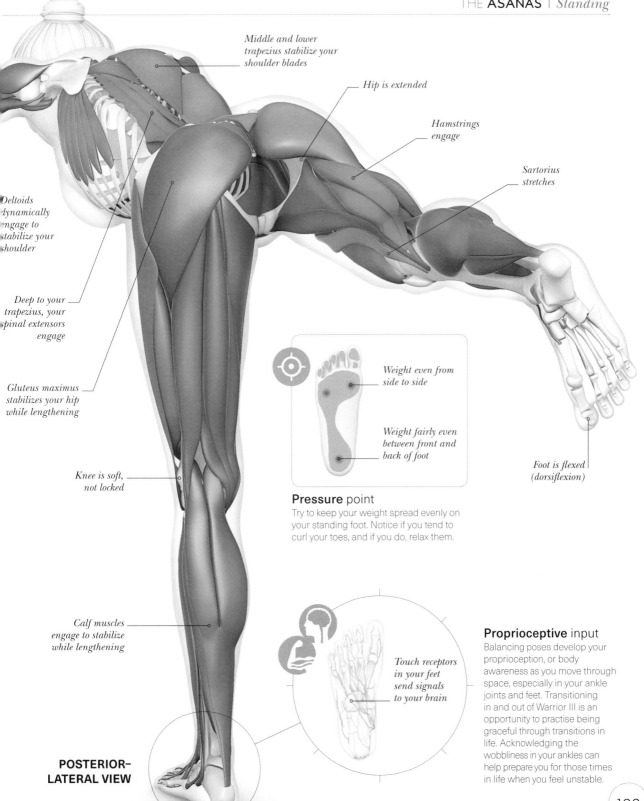

Middle and lower trapezius stabilize your shoulder blades

Hip is extended

Hamstrings engage

Sartorius stretches

Deltoids dynamically engage to stabilize your shoulder

Deep to your trapezius, your spinal extensors engage

Gluteus maximus stabilizes your hip while lengthening

Weight even from side to side

Weight fairly even between front and back of foot

Foot is flexed (dorsiflexion)

Knee is soft, not locked

Pressure point

Try to keep your weight spread evenly on your standing foot. Notice if you tend to curl your toes, and if you do, relax them.

Calf muscles engage to stabilize while lengthening

Touch receptors in your feet send signals to your brain

POSTERIOR-LATERAL VIEW

Proprioceptive input

Balancing poses develop your proprioception, or body awareness as you move through space, especially in your ankle joints and feet. Transitioning in and out of Warrior III is an opportunity to practise being graceful through transitions in life. Acknowledging the wobbliness in your ankles can help prepare you for those times in life when you feel unstable.

109

TREE
Vrksasana

Tree pose builds static balance, which can be facilitated by allowing a smooth and steady breath and focused mind. In this iconic yoga pose, unsteadiness is completely natural. Wobbling means you are strengthening muscles key for joint stabilization.

— *Arms raised over head*

VARIATION
Raising your arms overhead shifts your centre of gravity higher. Challenge your balance further by lifting your gaze. You may also hold your arms in a wide V.

THE BIG PICTURE

Large muscles in your standing thigh and lower leg engage to give your body a stable base. Muscles in your torso and on your raised thigh work to keep your leg lifted and rotated outwards. Your upper body remains neutral and stable.

Arms

Your **brachialis**, **biceps**, and **brachioradialis** flex your elbows while your **pectoralis major** helps to adduct your shoulders. Your **wrist flexors** stretch, while your **wrist extensors** engage as you firmly press your palms together at your sternum.

Shoulder
Pectoralis major
Triceps brachii
Elbow
Wrist
Biceps brachii
Brachialis

Rhomboids
Spinal extensors
Spine

Rectus abdominis
Transversus abdominis

Torso

Your **spinal extensors** and **transversus abdominis** engage to elongate and stabilize your spine into its neutral curves. Your **rhomboids** and **middle** and **lower trapezius** engage to retract your scapulae.

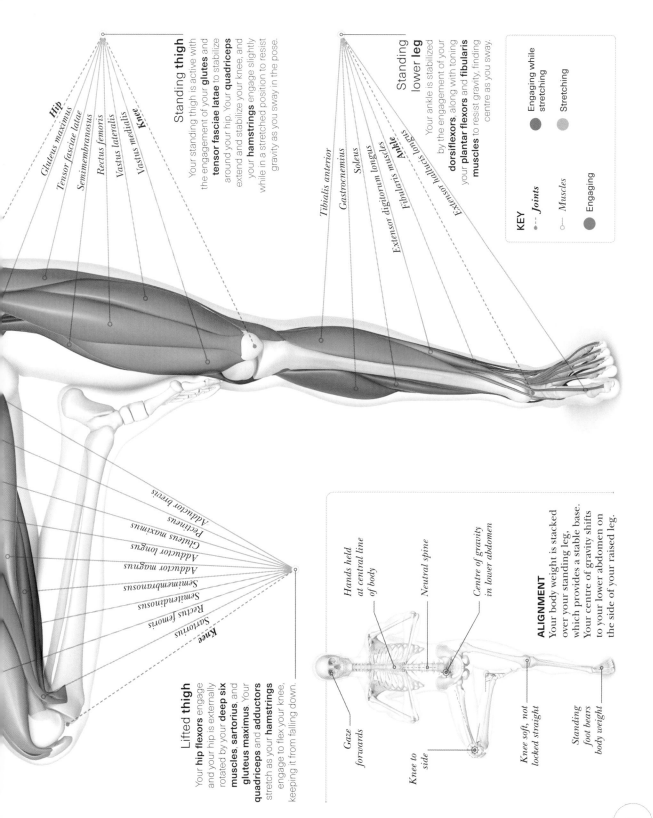

Standing thigh

Your standing thigh that is active with the engagement of your **glutes** and **tensor fasciae latae** to stabilize around your hip. Your **quadriceps** extend and stabilize your knee, and your **hamstrings** engage slightly while in a stretched position to resist gravity as you sway in the pose.

Hip

Gluteus maximus
Tensor fasciae latae
Semimembranosus
Rectus femoris
Vastus lateralis
Vastus medialis

Knee

Standing lower leg

Your ankle is stabilized by the engagement of your **dorsiflexors**, along with toning your **plantar flexors** and **fibularis muscles** to resist gravity, finding centre as you sway.

Tibialis anterior
Gastrocnemius
Soleus
Extensor digitorum longus

Ankle

Fibularis longus
Extensor hallucis longus

KEY
--- **Joints**
○— **Muscles**
● **Engaging**
● Engaging while stretching
● Stretching

Lifted thigh

Your **hip flexors** engage and your hip is externally rotated by your **deep six muscles**, **sartorius**, and **gluteus maximus**. Your **quadriceps** and **adductors** stretch as your **hamstrings** engage to flex your knee, keeping it from falling down.

Knee

Sartorius
Rectus femoris
Semitendinosus
Semimembranosus
Adductor magnus
Adductor longus
Gluteus maximus
Pectineus
Adductor brevis

ALIGNMENT

Your body weight is stacked over your standing leg, which provides a stable base. Your centre of gravity shifts to your lower abdomen on the side of your raised leg.

Hands held at central line of body

Neutral spine

Centre of gravity in lower abdomen

Gaze forwards

Knee to side

Knee soft, not locked straight

Standing foot bears body weight

⟫ CLOSER LOOK

Tree pose stabilizes your hips in a unique position. Holding the pose increases body awareness, particularly in the sole of your standing foot. Breathe steadily and focus.

Hip abductors

If you are not engaging your hip abductors, particularly your gluteus medius, on your standing thigh, your hip will hike out. This is tough for balance and a common bad postural habit you may mindlessly do. To counteract this, press your standing hip in, bringing your pelvis to neutral.

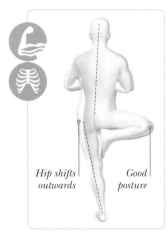

Hip shifts outwards   *Good posture*

Upper trapezius relaxes

Biceps brac engage as hands are at the hear

Spinal extensors engage to maintain posture

Quadriceps lengthen as knee flexes

Glutes engage strongly to keep hips aligned

Foot and thigh press into each other with equal and opposite force

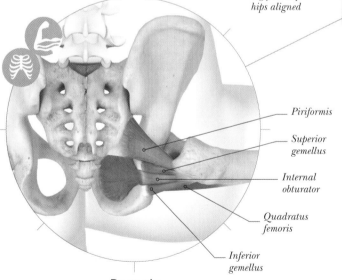

Piriformis

Superior gemellus

Internal obturator

Quadratus femoris

Inferior gemellus

Hamstrings engage to maintain balance

Deep six

To rotate your hip out to the side you engage a set of six small muscles deep within your hip joint. Strong standing poses, such as Tree, dynamically stretch and strengthen the deep six external rotators. To get a deeper stretch in these muscles, try stretches like King Pigeon (see pp.80–83).

Ligaments around your ankles help to stabilize

POSTERIOR VIEW

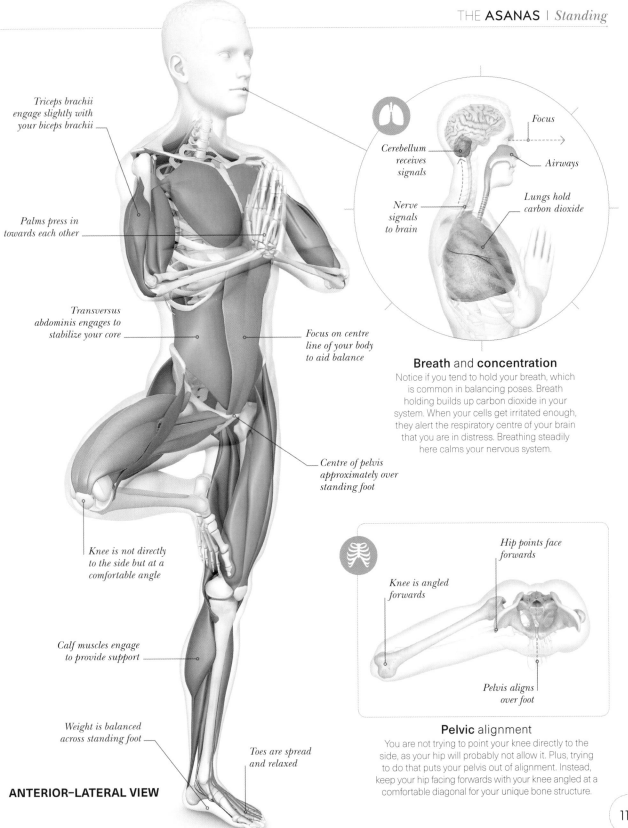

*Triceps brachii
engage slightly with
your biceps brachii*

*Palms press in
towards each other*

*Transversus
abdominis engages to
stabilize your core*

*Focus on centre
line of your body
to aid balance*

*Knee is not directly
to the side but at a
comfortable angle*

*Calf muscles engage
to provide support*

*Centre of pelvis
approximately over
standing foot*

*Weight is balanced
across standing foot*

*Toes are spread
and relaxed*

ANTERIOR-LATERAL VIEW

Focus

*Cerebellum
receives
signals*

Airways

*Nerve
signals
to brain*

*Lungs hold
carbon dioxide*

Breath and concentration

Notice if you tend to hold your breath, which
is common in balancing poses. Breath
holding builds up carbon dioxide in your
system. When your cells get irritated enough,
they alert the respiratory centre of your brain
that you are in distress. Breathing steadily
here calms your nervous system.

*Hip points face
forwards*

*Knee is angled
forwards*

*Pelvis aligns
over foot*

Pelvic alignment

You are not trying to point your knee directly to the
side, as your hip will probably not allow it. Plus, trying
to do that puts your pelvis out of alignment. Instead,
keep your hip facing forwards with your knee angled at a
comfortable diagonal for your unique bone structure.

DANCER

Natarajasana

Dancer pose is a challenging static balancing pose, which also develops strength, flexibility, and agility. Dynamic balance skills are required to transition in and out of the pose with grace, though you can always hold onto a wall or chair for steadiness.

THE BIG PICTURE

Large muscles of your standing hip, thigh, and leg dramatically engage to help you balance on one leg. The front of your lifted hip and thigh stretch, as you kick back as a counterbalance. Your back muscles engage to come into a backbend, while your chest and abdomen stretch. Your neck is extended out long, and your shoulders are relaxed.

KEY
- ●--- *Joints*
- ○— *Muscles*
- ● Engaging
- ● Engaging while stretching
- ● Stretching

Arms

In your front arm, your **anterior deltoid, pectoralis major,** and **coracobrachialis** flex your shoulder, while your **triceps** extend your elbow. In your back arm, your **posterior deltoid, latissimus dorsi,** and **teres major** engage to extend your shoulder, while your **triceps** extend your elbow. Your **elbow flexors** also engage in a stretched position to isometrically pull your leg inward.

Brachioradialis
Elbow
Biceps brachii
Triceps brachii
Deltoids
Shoulder
Pectoralis major

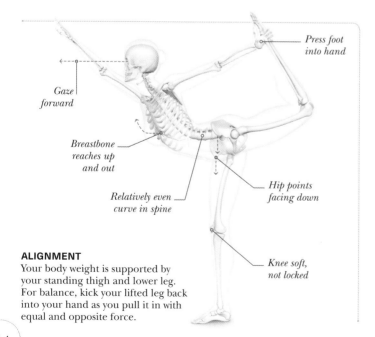

Press foot into hand

Gaze forward

Breastbone reaches up and out

Relatively even curve in spine

Hip points facing down

Knee soft, not locked

ALIGNMENT
Your body weight is supported by your standing thigh and lower leg. For balance, kick your lifted leg back into your hand as you pull it in with equal and opposite force.

Arms reach up and back

Both hands hold foot

VARIATION
For a challenge, reach both arms up and back to grasp your big toe. If you feel pinching in your lower back, don't go as deeply into the bend. Also try using a strap looped around your ankle.

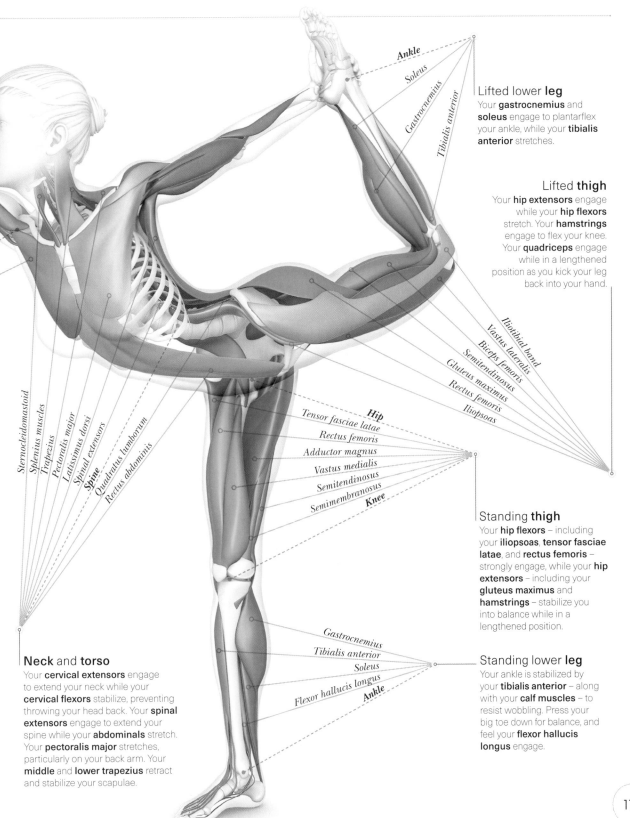

Ankle

Soleus

Gastrocnemius

Tibialis anterior

Lifted lower **leg**
Your **gastrocnemius** and **soleus** engage to plantarflex your ankle, while your **tibialis anterior** stretches.

Lifted **thigh**
Your **hip extensors** engage while your **hip flexors** stretch. Your **hamstrings** engage to flex your knee. Your **quadriceps** engage while in a lengthened position as you kick your leg back into your hand.

Iliotibial band

Vastus lateralis

Biceps femoris

Semitendinosus

Gluteus maximus

Rectus femoris

Iliopsoas

Sternocleidomastoid

Splenius muscles

Trapezius

Pectoralis major

Latissimus dorsi

Spinal extensors

Spine

Quadratus lumborum

Rectus abdominis

Hip

Tensor fasciae latae

Rectus femoris

Adductor magnus

Vastus medialis

Semitendinosus

Semimembranosus

Knee

Standing **thigh**
Your **hip flexors** – including your **iliopsoas**, **tensor fasciae latae**, and **rectus femoris** – strongly engage, while your **hip extensors** – including your **gluteus maximus** and **hamstrings** – stabilize you into balance while in a lengthened position.

Gastrocnemius

Tibialis anterior

Soleus

Flexor hallucis longus

Ankle

Neck and **torso**
Your **cervical extensors** engage to extend your neck while your **cervical flexors** stabilize, preventing throwing your head back. Your **spinal extensors** engage to extend your spine while your **abdominals** stretch. Your **pectoralis major** stretches, particularly on your back arm. Your **middle** and **lower trapezius** retract and stabilize your scapulae.

Standing lower **leg**
Your ankle is stabilized by your **tibialis anterior** – along with your **calf muscles** – to resist wobbling. Press your big toe down for balance, and feel your **flexor hallucis longus** engage.

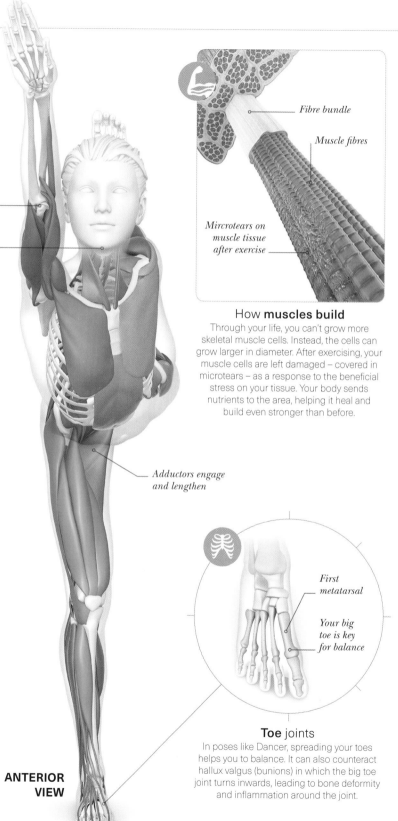

»CLOSER LOOK

Dancer strikes a balance between
stability and mobility, along with
effort and ease. Muscles build in
strengthening poses like this
when microtears heal.

Elbow soft,
not locked

Neck muscles stabilize
chin, lifting slightly

Fibre bundle

Muscle fibres

Mircotears on
muscle tissue
after exercise

How **muscles build**

Through your life, you can't grow more
skeletal muscle cells. Instead, the cells can
grow larger in diameter. After exercising, your
muscle cells are left damaged – covered in
microtears – as a response to the beneficial
stress on your tissue. Your body sends
nutrients to the area, helping it heal and
build even stronger than before.

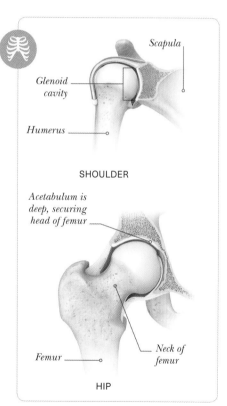

Scapula

Glenoid
cavity

Humerus

SHOULDER

Acetabulum is
deep, securing
head of femur

Femur

Neck of
femur

HIP

Adductors engage
and lengthen

First
metatarsal

Your big
toe is key
for balance

ANTERIOR
VIEW

Ball and **socket** joints

Both your shoulder and hip joints are ball and
socket joints (see pp.16–17). Your shoulder joint
is shallow with a lot of mobility; it is only limited
by ligaments and muscles. Your hip joint, in
contrast, is deeper with more joint structures
to help hold it securely in place.

Toe joints

In poses like Dancer, spreading your toes
helps you to balance. It can also counteract
hallux valgus (bunions) in which the big toe
joint turns inwards, leading to bone deformity
and inflammation around the joint.

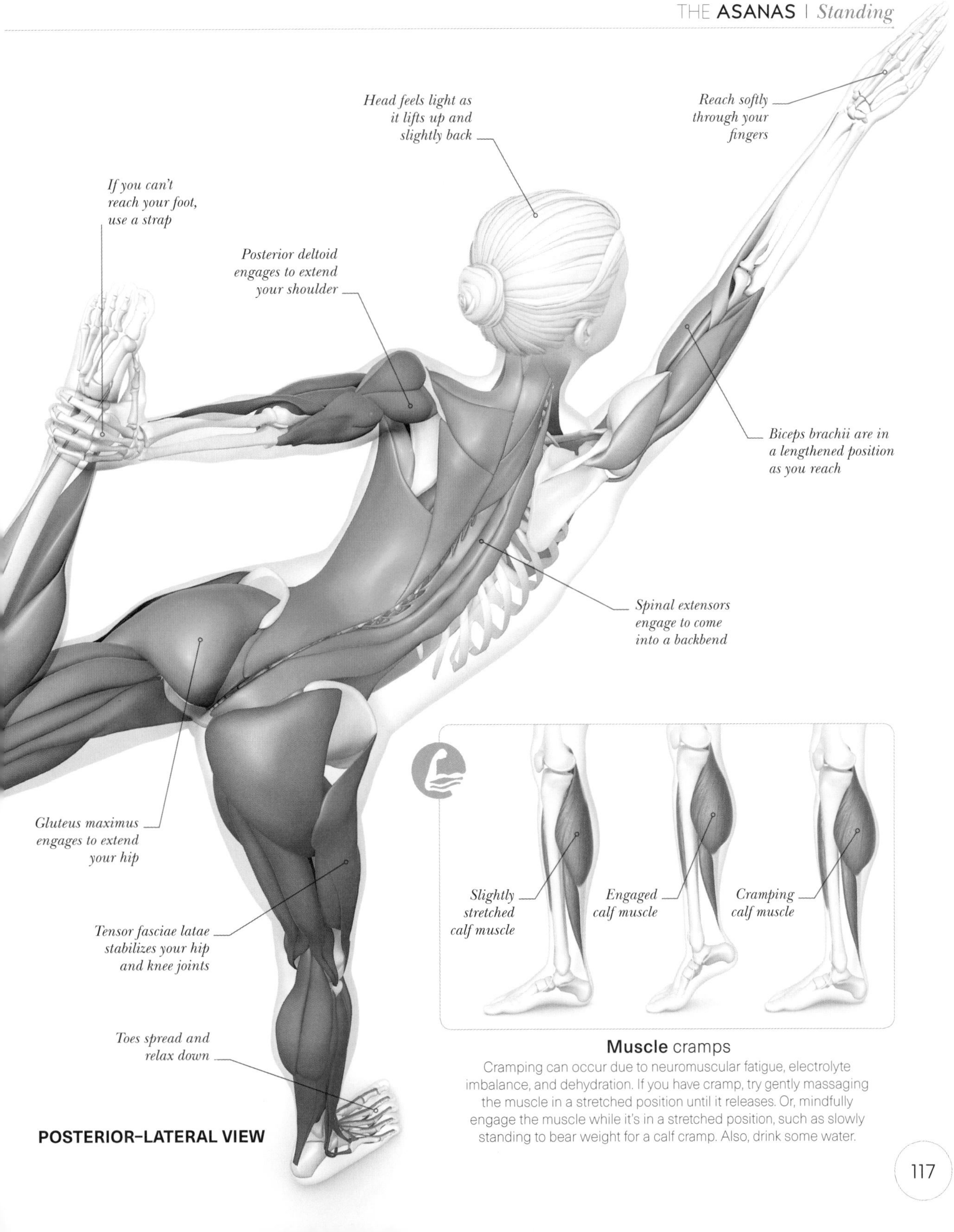

*Head feels light as
it lifts up and
slightly back*

*Reach softly
through your
fingers*

*If you can't
reach your foot,
use a strap*

*Posterior deltoid
engages to extend
your shoulder*

*Biceps brachii are in
a lengthened position
as you reach*

*Spinal extensors
engage to come
into a backbend*

*Gluteus maximus
engages to extend
your hip*

*Tensor fasciae latae
stabilizes your hip
and knee joints*

*Toes spread and
relax down*

POSTERIOR–LATERAL VIEW

*Slightly
stretched
calf muscle*

*Engaged
calf muscle*

*Cramping
calf muscle*

Muscle cramps

Cramping can occur due to neuromuscular fatigue, electrolyte imbalance, and dehydration. If you have cramp, try gently massaging the muscle in a stretched position until it releases. Or, mindfully engage the muscle while it's in a stretched position, such as slowly standing to bear weight for a calf cramp. Also, drink some water.

117

TRIANGLE

Trikonasana

Triangle is a strengthening and grounding standing pose. It involves twisting your spine and ribcage to move against gravity and the tendency to round forward and down. Strong poses like this can strengthen both your muscles and bones.

THE BIG PICTURE

This pose particularly strengthens your core, thighs, and legs. Deep muscles close to your spine engage to stabilize your spine and give your brain feedback, enhancing your mind–body connection.

Neck and **torso**

To rotate your neck, on the side nearer the ground (model's left), your **sternocleidomastoid, rotatores, multifidus,** and **semispinalis cervicis** engage, while stretching on the upward-facing side (model's right). On the upward side, your **splenius capitis** and **splenius cervicis** engage, while stretching on the downward side. Your **transversus abdominis** engages to stabilize your spine. On the upward-facing side, your **external obliques** stretch, while your **internal obliques** engage to rotate your spine. On the downward side, your **external obliques** engage to rotate your spine.

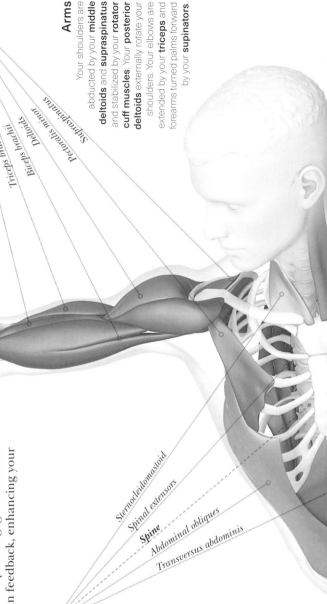

ALIGNMENT
Rotate your back hip inwards by turning your toes inwards. Rotate your front hip outwards by turning your toes towards the front of your mat. Rotate your spine to stack your shoulder blades vertically.

Arm reaches up

Bring shoulder blade back

Back hip rotating inwards

Front hip rotating outwards

Knees soft, not locked

Centre of gravity

Press into outer edge of foot

Supinator

Elbow

Brachialis

Triceps brachii

Biceps brachii

Deltoids

Pectoralis minor

Supraspinatus

Arms
Your shoulders are abducted by your **middle deltoids** and **supraspinatus** and stabilized by your **rotator cuff muscles**. Your **posterior deltoids** externally rotate your shoulders. Your elbows are extended by your **triceps** and forearms turned palms forward by your **supinators**.

Sternocleidomastoid

Spinal extensors

Spine

Abdominal obliques

Transversus abdominis

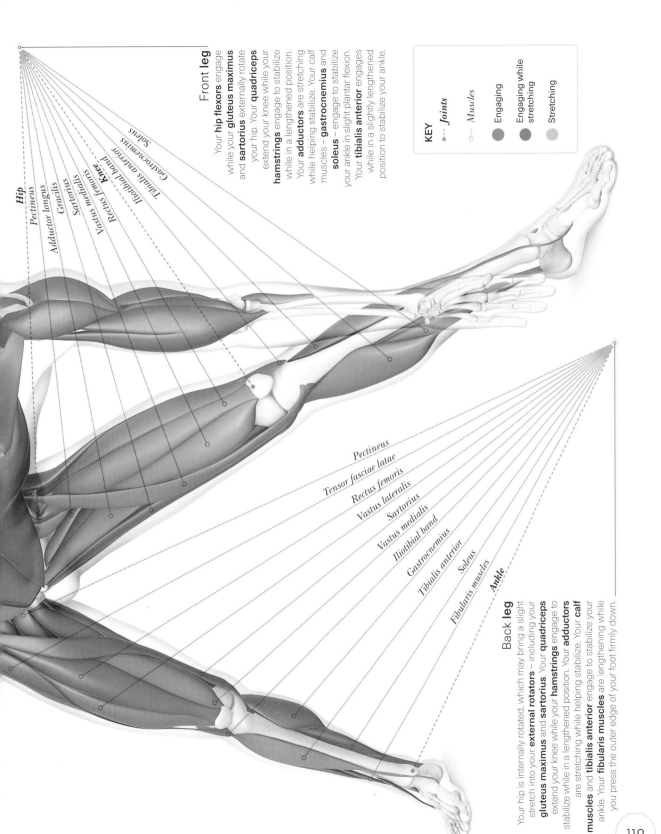

Front leg

Your **hip flexors** engage while your **gluteus maximus** and **sartorius** externally rotate your hip. Your **quadriceps** extend your knee while your **hamstrings** engage to stabilize while in a lengthened position. Your **adductors** are stretching while helping stabilize. Your calf muscles – **gastrocnemius** and **soleus** – engage to stabilize your ankle in slight plantar flexion. Your **tibialis anterior** engages while in a slightly lengthened position to stabilize your ankle.

KEY

- •-- *Joints*
- ○-- *Muscles*
- ● Engaging
- ● Engaging while stretching
- ● Stretching

Hip
Pectineus
Adductor longus
Gracilis
Sartorius
Vastus medialis
Rectus femoris
Knee
Iliotibial band
Tibialis anterior
Gastrocnemius
Soleus

Pectineus
Tensor fasciae latae
Rectus femoris
Vastus lateralis
Sartorius
Vastus medialis
Iliotibial band
Gastrocnemius
Tibialis anterior
Soleus
Fibularis muscles
Ankle

Back leg

Your hip is internally rotated, which may bring a slight stretch into your **external rotators** – including your **gluteus maximus** and **sartorius**. Your **quadriceps** extend your knee while your **hamstrings** engage to stabilize while in a lengthened position. Your **adductors** are stretching while helping stabilize. Your **calf muscles** and **tibialis anterior** engage to stabilize your ankle. Your **fibularis muscles** are lengthening while you press the outer edge of your foot firmly down.

119

TRIANGLE | *Trikonasana*

» CLOSER LOOK

Strengthening the muscles of your thighs, hips and back, in poses like Triangle, may have the added benefit of boosting bone density. This pose should be practised with care, listen to your body and ease out of the pose if you experience any pain or tingling; and be mindful of your knee joints.

Pressure points

Ease off or come out of any pose that causes numbness, sharpness, or shooting pain. This may be due to pressure or impingement on nerves. Likewise, stop if you experience any tingling, coolness or a dull, lifeless feeling like when you fall asleep on your arm. This can be caused by pressure occluding blood vessels.

Brachial nerve plexus

Scalene muscles may press on nerves

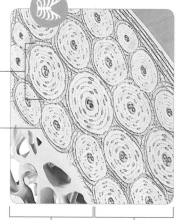

Osteoblasts at edge of compact bone *Osteon*

Spongy bone

Compact bone

Bone growth

Large muscles in your thighs engage firmly, beneficially stressing your bones. This may wake up cells in the bone called osteoblasts, which triggers bone building. A 10-year trial concluded that yoga appears to raise bone mineral density in the spine and the femur.

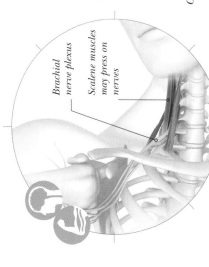

Transversospinales (including the rotatores and multifidus) engage to rotate with awareness

Rotator cuff muscles stabilize the shoulders

Neck muscles dynamically

VARIATION

Twisted Triangle adds a torso twist to the pose, which challenges your stability. With your right foot forwards, reach over your front leg and rotate your torso to the right. Avoid this pose if you have back issues. Feel free to place your left hand on your leg, a block, or the floor.

Torso twists upwards

Press outer edge of back heel down

Hand reaches down

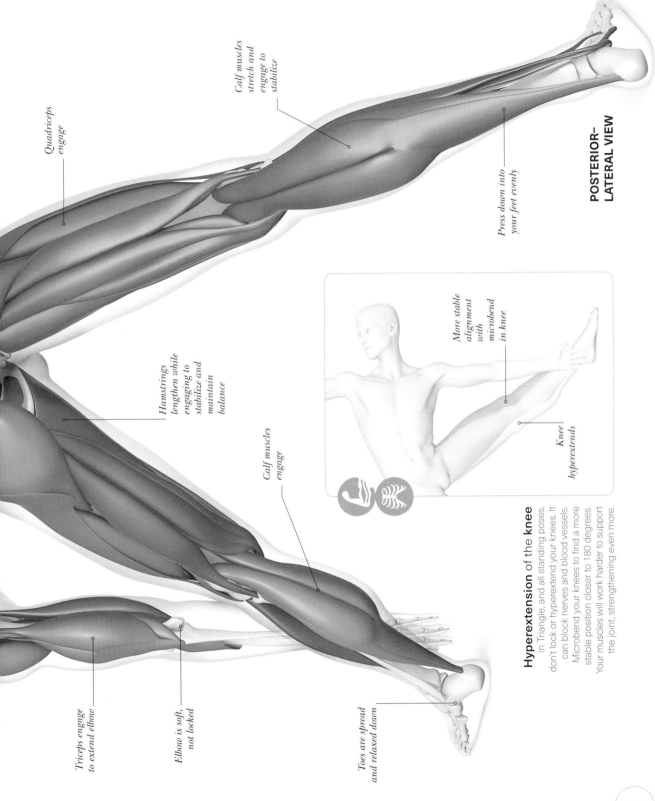

Quadriceps engage

Calf muscles stretch and engage to stabilize

Press down into your feet evenly

Hamstrings lengthen while engaging to stabilize and maintain balance

Calf muscles engage

More stable alignment with microbend in knee

Knee hyperextends

Triceps engage to extend elbow

Elbow is soft, not locked

Toes are spread and relaxed down

Hyperextension of the knee

In Triangle, and all standing poses, don't lock or hyperextend your knees. It can block nerves and blood vessels. Microbend your knees to find a more stable position closer to 180 degrees. Your muscles will work harder to support the joint, strengthening even more.

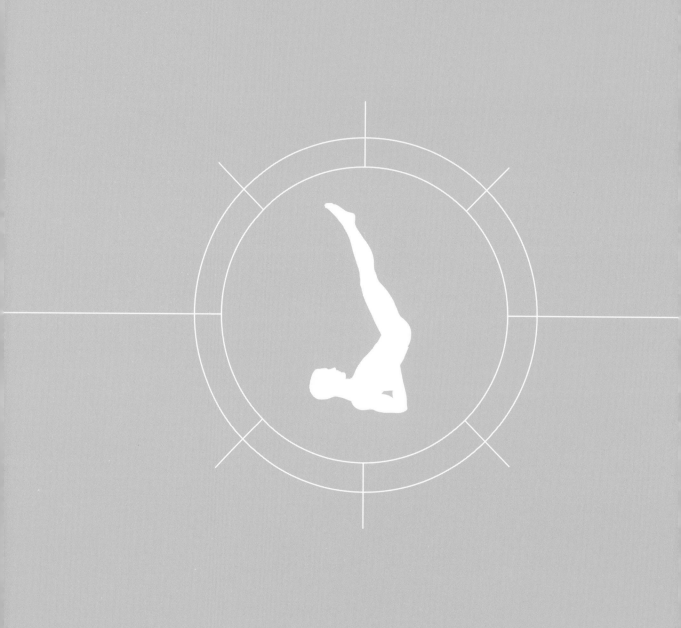

INVERSION
ASANAS

Inversions are defined here as poses that turn your body upside
down. Having your head below your heart has certain physiological
effects and benefits, such as boosting circulation and aiding
lymphatic drainage. Full inversions can be a great exploration
of getting a new perspective – both literally and figuratively.

DOWNWARD-FACING DOG

Adho Mukha Svanasana

Also known as "Down Dog", this is a common pose in modern yoga classes, particularly as an integral part of Sun salutations or flow sequences. This arm balance is a forward fold and partial inversion, stretching the back of your legs and strengthening your shoulders.

THE BIG PICTURE

In this pose, the back of your body – including your buttocks, thighs, and calf muscles – is stretching. Your shoulders are strengthening as you press into the floor.

Torso

Your **transversus abdominis** stabilizes your spine and core. Your **spinal extensors** engage while your spine remains neutral or in slight extension. Your **middle** and **lower trapezius** engage to stabilize and slightly depress your scapulae. Your **latissimus dorsi** stretches.

Transversus abdominis
Spine
Rectus abdominis
Serratus anterior
Latissimus dorsi
Pectoralis major
Trapezius
Infraspinatus
Teres minor

Arms

Your **shoulder flexors** engage – including your **pectoralis major**, which has some lengthening muscle fibres due to shoulder external rotation and slight abduction. Your **deltoids** dynamically engage to stabilize your shoulder in position, and externally rotate your shoulders with the help of your **infraspinatus** and **teres minor**. Your **rotator cuff muscles** are active to stabilize your shoulders. Your **triceps** extend your elbows.

Shoulder
Deltoids
Triceps brachii
Biceps brachii
Elbow
Pronator teres
Brachioradialis
Wrist
Pronator quadratus

Splenius muscles

Neck

Your **splenius capitis, splenius cervicis,** and **upper trapezius** are either fully relaxed and stretching, or slightly engaging while lengthening to keep your ears approximately in line with your arms.

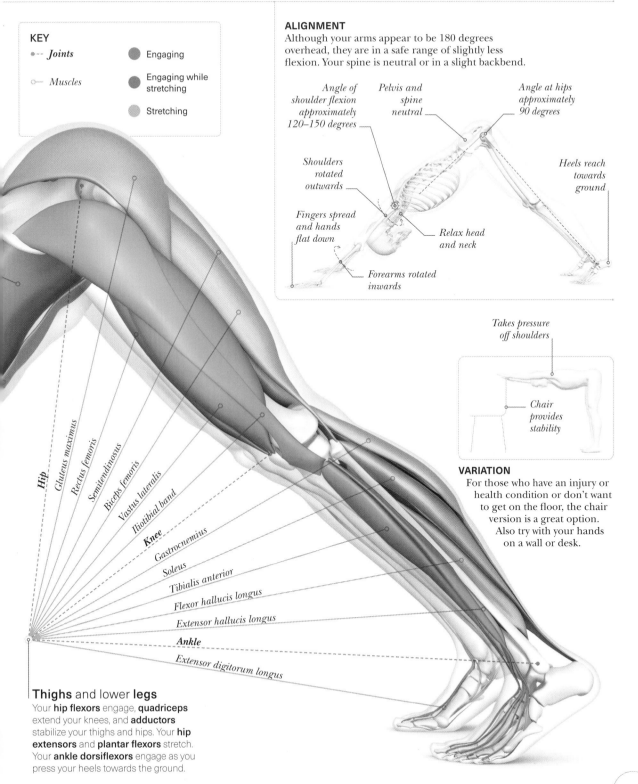

KEY

●--- *Joints*

○— *Muscles*

● Engaging

● Engaging while stretching

● Stretching

ALIGNMENT

Although your arms appear to be 180 degrees overhead, they are in a safe range of slightly less flexion. Your spine is neutral or in a slight backbend.

Angle of shoulder flexion approximately 120–150 degrees

Pelvis and spine neutral

Angle at hips approximately 90 degrees

Shoulders rotated outwards

Heels reach towards ground

Fingers spread and hands flat down

Relax head and neck

Forearms rotated inwards

Takes pressure off shoulders

Chair provides stability

VARIATION

For those who have an injury or health condition or don't want to get on the floor, the chair version is a great option. Also try with your hands on a wall or desk.

Hip

Gluteus maximus

Rectus femoris

Semitendinosus

Biceps femoris

Vastus lateralis

Iliotibial band

Knee

Gastrocnemius

Soleus

Tibialis anterior

Flexor hallucis longus

Extensor hallucis longus

Ankle

Extensor digitorum longus

Thighs and lower legs

Your **hip flexors** engage, **quadriceps** extend your knees, and **adductors** stabilize your thighs and hips. Your **hip extensors** and **plantar flexors** stretch. Your **ankle dorsiflexors** engage as you press your heels towards the ground.

›› CLOSER LOOK

Being too tight or too flexible can both present challenges when finding effective alignment in Downward-facing Dog. However, modifications can make it accessible for everyone.

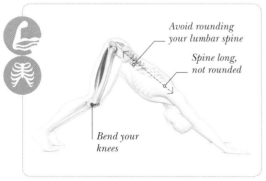

Avoid rounding your lumbar spine

Spine long, not rounded

Bend your knees

Tight **hamstrings**

When the hamstrings are tight, the pelvis is pulled and the back rounded. The integrity of your spine is more important than having your legs straight in this pose, so flex your knees and press into the floor, helping to lengthen your spine and bring the pelvis closer to neutral.

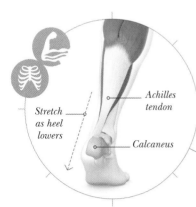

Stretch as heel lowers

Achilles tendon

Calcaneus

Achilles tendon

This tendon was named after the Greek mythological figure, Achilles, who only had one weakness: his calcaneal tendon. For many, it is very tight, preventing touching the heel to the ground in this pose. It can stretch with practice; however, there is some functional benefit to it maintaining tension as its stores potential energy.

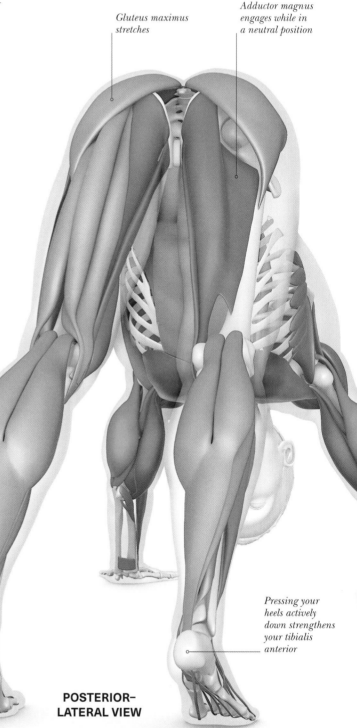

Gluteus maximus stretches

Adductor magnus engages while in a neutral position

Pressing your heels actively down strengthens your tibialis anterior

POSTERIOR-LATERAL VIEW

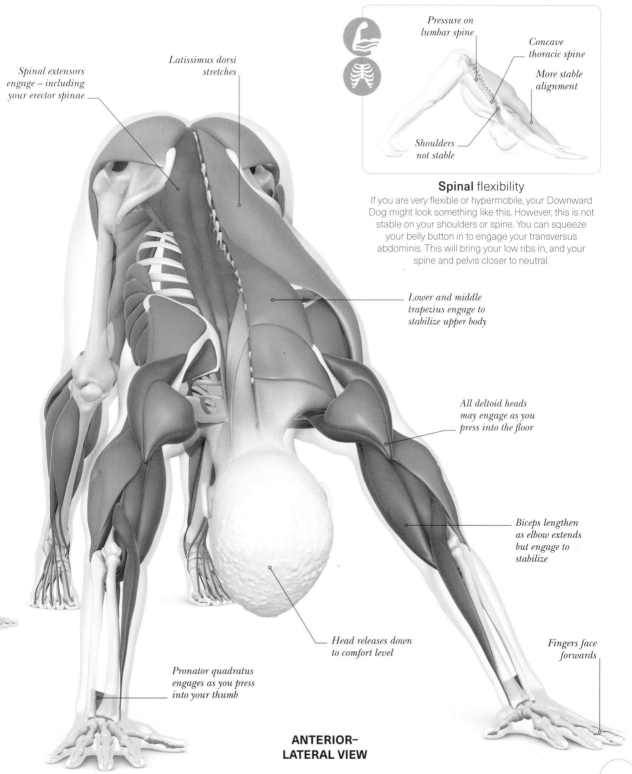

Spinal extensors engage – including your erector spinae

Latissimus dorsi stretches

Pressure on lumbar spine

Concave thoracic spine

More stable alignment

Shoulders not stable

Spinal flexibility

If you are very flexible or hypermobile, your Downward Dog might look something like this. However, this is not stable on your shoulders or spine. You can squeeze your belly button in to engage your transversus abdominis. This will bring your low ribs in, and your spine and pelvis closer to neutral.

Lower and middle trapezius engage to stabilize upper body

All deltoid heads may engage as you press into the floor

Biceps lengthen as elbow extends but engage to stabilize

Head releases down to comfort level

Fingers face forwards

Pronator quadratus engages as you press into your thumb

ANTERIOR–LATERAL VIEW

HEADSTAND

Sirsasana

This full inversion turns you physically upside-down. A multitude of benefits can be ascribed to this pose: from helping you to breathe more efficiently to strengthening your upper body – especially the muscles around your shoulder joints – and your core.

THE BIG PICTURE

This pose strengthens your arms and shoulders. Your core and thighs activate to stabilize your body at its centre, preventing you from falling to either side. Despite the name, it is your arms that are supporting your weight in this pose, not your head.

KEY

- •--- *Joints*
- •— *Muscles*
- ⬤ Engaging
- ⬤ Engaging while stretching
- ⬤ Stretching

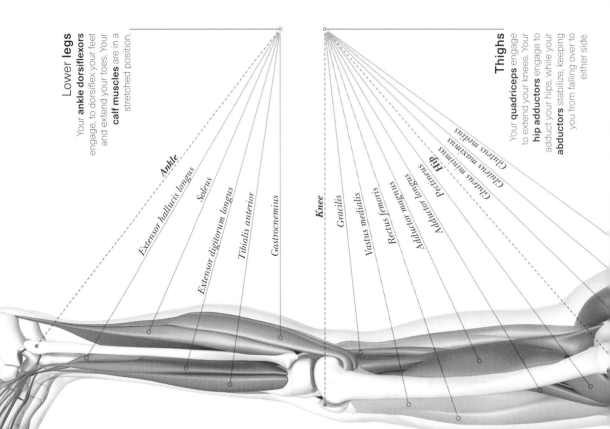

Lower legs
Your **ankle dorsiflexors** engage, to dorsiflex your feet and extend your toes. Your **calf muscles** are in a stretched position.

Ankle

Extensor hallucis longus

Soleus

Extensor digitorum longus

Tibialis anterior

Gastrocnemius

Knee

Gracilis

Vastus medialis

Rectus femoris

Adductor magnus

Adductor longus

Hip

Pectineus

Gluteus minimus

Gluteus maximus

Gluteus medius

Thighs
Your **quadriceps** engage to extend your knees. Your **hip adductors** engage to adduct your hips, while your **abductors** stabilize, keeping you from falling over to either side.

VARIATION
This version of the pose has a reduced risk of falling and takes weight off your upper body. Push your forearms into the floor, lower your heels, and lift hips up and back. Allow your head to drop.

Feet hip distance apart, heels reach down

Forearms support upper body weight

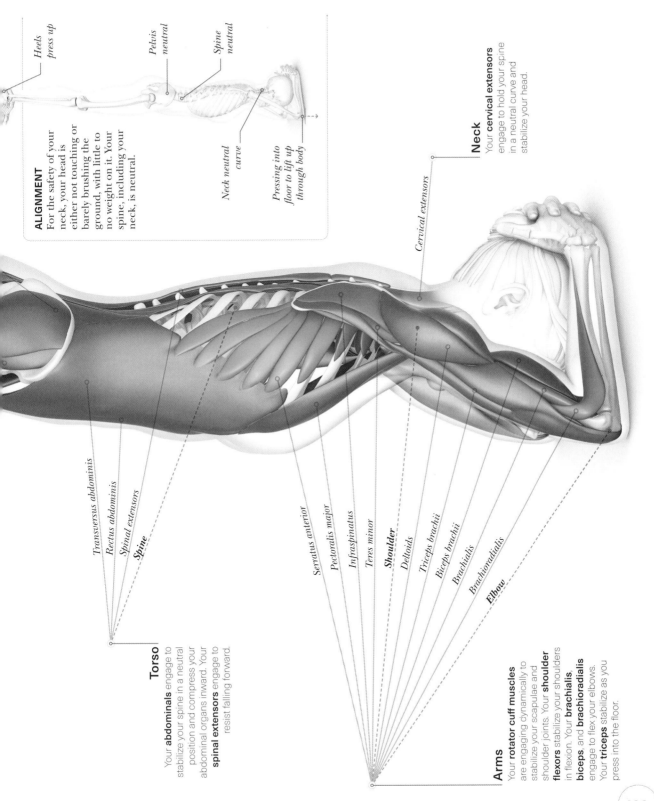

ALIGNMENT

For the safety of your neck, your head is either not touching or barely brushing the ground, with little to no weight on it. Your spine, including your neck, is neutral.

Heels press up

Pelvis neutral

Spine neutral

Neck neutral curve

Pressing into floor to lift up through body

Neck

Your **cervical extensors** engage to hold your spine in a neutral curve and stabilize your head.

Cervical extensors

Transversus abdominis

Rectus abdominis

Spinal extensors

Spine

Serratus anterior

Pectoralis major

Infraspinatus

Teres minor

Shoulder

Deltoids

Triceps brachii

Biceps brachii

Brachialis

Brachioradialis

Elbow

Torso

Your **abdominals** engage to stabilize your spine in a neutral position and compress your abdominal organs inward. Your **spinal extensors** engage to resist falling forward.

Arms

Your **rotator cuff muscles** are engaging dynamically to stabilize your scapulae and shoulder joints. Your **shoulder flexors** stabilize your shoulders in flexion. Your **brachialis, biceps,** and **brachioradialis** engage to flex your elbows. Your **triceps** stabilize as you press into the floor.

129

» CLOSER LOOK

Headstand can be safely practised with little to no pressure on the head and neck. It has many health benefits, from improving respiratory and shoulder function to helping you better regulate your blood pressure.

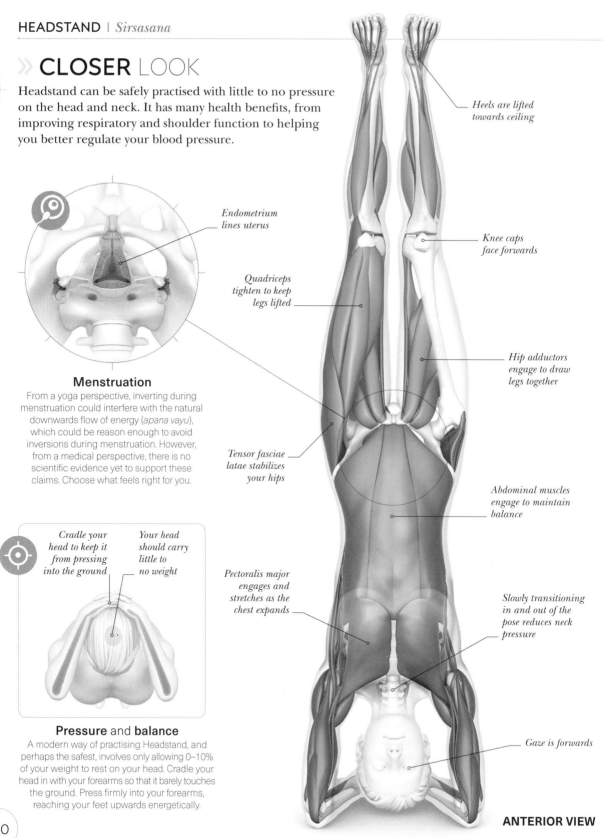

Endometrium lines uterus

Menstruation

From a yoga perspective, inverting during menstruation could interfere with the natural downwards flow of energy (*apana vayu*), which could be reason enough to avoid inversions during menstruation. However, from a medical perspective, there is no scientific evidence yet to support these claims. Choose what feels right for you.

Cradle your head to keep it from pressing into the ground

Your head should carry little to no weight

Pressure and balance

A modern way of practising Headstand, and perhaps the safest, involves only allowing 0–10% of your weight to rest on your head. Cradle your head in with your forearms so that it barely touches the ground. Press firmly into your forearms, reaching your feet upwards energetically.

Heels are lifted towards ceiling

Knee caps face forwards

Quadriceps tighten to keep legs lifted

Hip adductors engage to draw legs together

Tensor fasciae latae stabilizes your hips

Abdominal muscles engage to maintain balance

Pectoralis major engages and stretches as the chest expands

Slowly transitioning in and out of the pose reduces neck pressure

Gaze is forwards

ANTERIOR VIEW

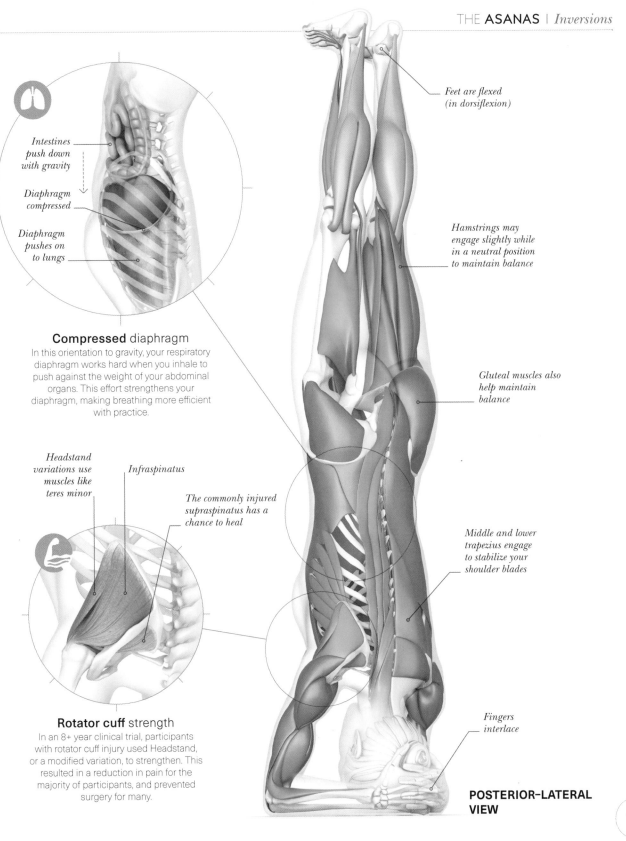

*Feet are flexed
(in dorsiflexion)*

*Hamstrings may
engage slightly while
in a neutral position
to maintain balance*

*Gluteal muscles also
help maintain
balance*

*Middle and lower
trapezius engage
to stabilize your
shoulder blades*

*Fingers
interlace*

**POSTERIOR–LATERAL
VIEW**

*Intestines
push down
with gravity*

*Diaphragm
compressed*

*Diaphragm
pushes on
to lungs*

Compressed diaphragm

In this orientation to gravity, your respiratory
diaphragm works hard when you inhale to
push against the weight of your abdominal
organs. This effort strengthens your
diaphragm, making breathing more efficient
with practice.

*Headstand
variations use
muscles like
teres minor*

Infraspinatus

*The commonly injured
supraspinatus has a
chance to heal*

Rotator cuff strength

In an 8+ year clinical trial, participants
with rotator cuff injury used Headstand,
or a modified variation, to strengthen. This
resulted in a reduction in pain for the
majority of participants, and prevented
surgery for many.

HALF SHOULDERSTAND

Ardha Sarvangasana

Shoulderstand is a classic inversion, often done at the end of an asana class to relax. It can help lower your blood pressure and activate the rest, digest, and rejuvenate part of your nervous system. The version shown here reduces pressure on the neck.

THE BIG PICTURE

This pose gently strengthens the muscles at the front of your neck, while your upper back and neck muscles stretch. The muscles of your core and thighs engage to stabilize you and hold your body in an inverted position.

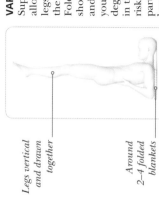

VARIATION

Supported shoulderstand allows you to bring your legs vertical to perform the traditional pose safely. Folded blankets under the shoulders take pressure and the sharp angle off your neck. Reducing the degree of neck flexion in this way lessens the risk of injury, particularly if you have neck issues.

Legs vertical and drawn together

Around 2–4 folded blankets

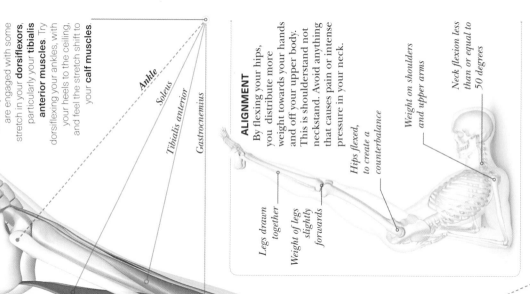

Lower legs

With your feet in plantar flexion, your **calf muscles** are engaged with some stretch in your **dorsiflexors**, particularly your **tibialis anterior muscles.** Try dorsiflexing your ankles, with your heels to the ceiling, and feel the stretch shift to your **calf muscles.**

Ankle

Soleus

Tibialis anterior

Gastrocnemius

ALIGNMENT

By flexing your hips, you distribute more weight towards your hands and off your upper body. This is shoulderstand not neckstand. Avoid anything that causes pain or intense pressure in your neck.

Legs drawn together

Weight of legs slightly forwards

Hips flexed, to create a counterbalance

Weight on shoulders and upper arms

Neck flexion less than or equal to 50 degrees

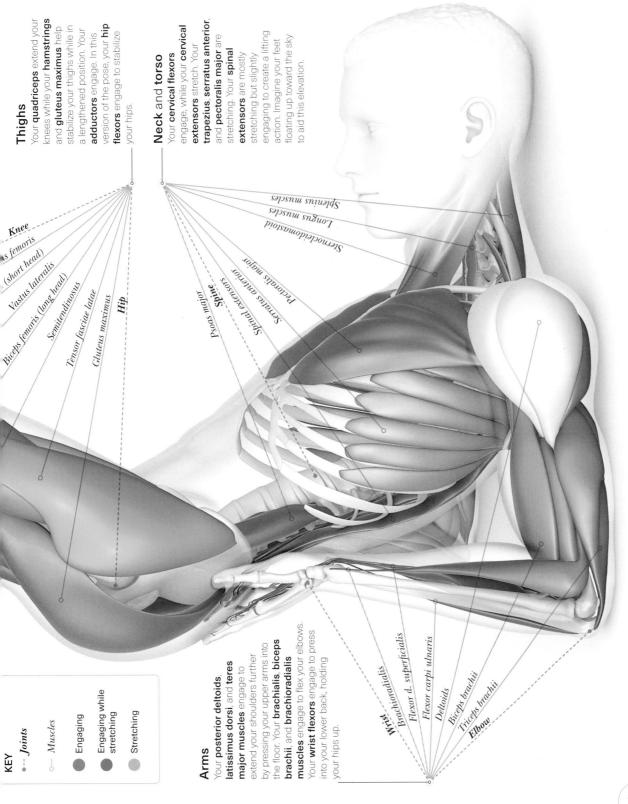

Thighs

Your **quadriceps** extend your knees while your **hamstrings** and **gluteus maximus** help stabilize your thighs while in a lengthened position. Your **adductors** engage. In this version of the pose, your **hip flexors** engage to stabilize your hips.

Neck and torso

Your **cervical flexors** engage, while your **cervical extensors** stretch. Your **trapezius**, **serratus anterior**, and **pectoralis major** are stretching. Your **spinal extensors** are mostly stretching but slightly engaging to create a lifting action. Imagine your feet floating up toward the sky to aid this elevation.

Splenius muscles

Longus muscles

Sternocleidomastoid

Pectoralis major

Spinal extensors

Spine

Psoas major

Knee

...s femoris

...(short head)

Vastus lateralis

Biceps femoris (long head)

Semitendinosus

Tensor fasciae latae

Gluteus maximus

Hip

Arms

Your **posterior deltoids**, **latissimus dorsi**, and **teres major muscles** engage to extend your shoulders further by pressing your upper arms into the floor. Your **brachialis**, **biceps brachii**, and **brachioradialis muscles** engage to flex your elbows. Your **wrist flexors** engage to press into your lower back, holding your hips up.

Wrist

Brachioradialis

Flexor d. superficialis

Flexor carpi ulnaris

Deltoids

Biceps brachii

Triceps brachii

Elbow

KEY

- --- *Joints*
- ○ *Muscles*
- ● Engaging
- ● Engaging while stretching
- ● Stretching

»» CLOSER LOOK

Shoulderstand is particularly effective at encouraging lymphatic drainage and improving overall circulation. Although it may not stimulate your thyroid, it can stimulate baroreceptors to lower your blood pressure.

Lymphatic drainage

Lymph vessels rely on movement to pump lymphatic fluid around your body. Like veins, they have one-way valves that prevent backflow (see opposite). Inverting encourages these valves to open, preventing or alleviating oedema (a build-up of fluids) in your ankles.

Lymph vessels

Lymph nodes

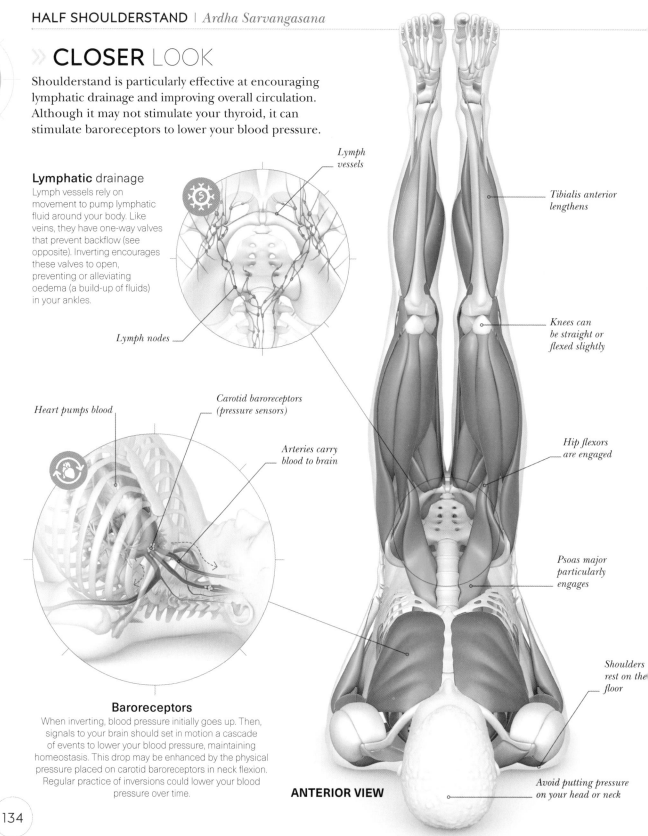

Tibialis anterior lengthens

Knees can be straight or flexed slightly

Hip flexors are engaged

Psoas major particularly engages

Shoulders rest on the floor

Heart pumps blood

Carotid baroreceptors (pressure sensors)

Arteries carry blood to brain

Baroreceptors

When inverting, blood pressure initially goes up. Then, signals to your brain should set in motion a cascade of events to lower your blood pressure, maintaining homeostasis. This drop may be enhanced by the physical pressure placed on carotid baroreceptors in neck flexion. Regular practice of inversions could lower your blood pressure over time.

ANTERIOR VIEW

Avoid putting pressure on your head or neck

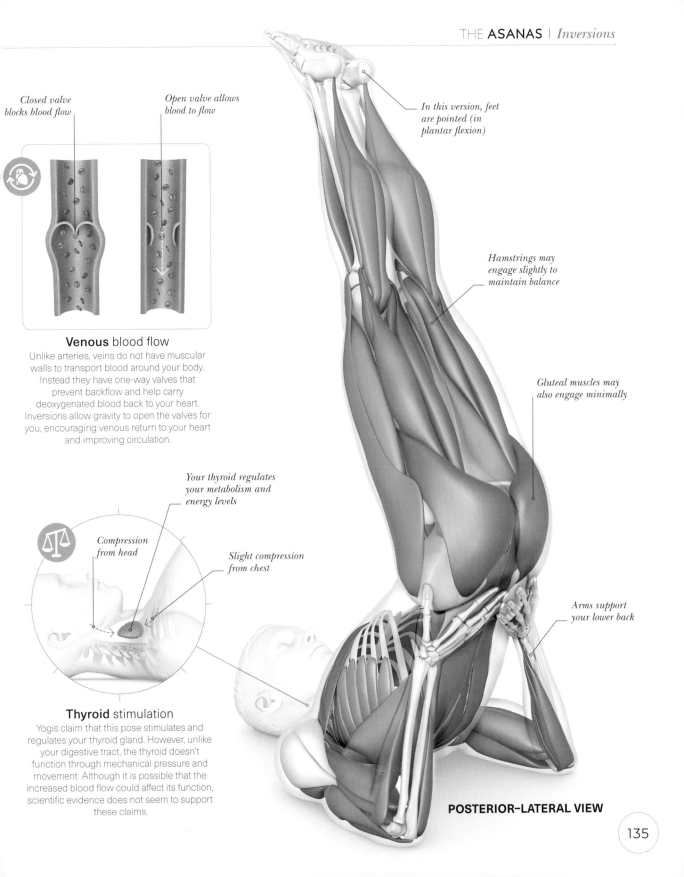

Closed valve blocks blood flow

Open valve allows blood to flow

Venous blood flow
Unlike arteries, veins do not have muscular walls to transport blood around your body. Instead they have one-way valves that prevent backflow and help carry deoxygenated blood back to your heart. Inversions allow gravity to open the valves for you, encouraging venous return to your heart and improving circulation.

Your thyroid regulates your metabolism and energy levels

Compression from head

Slight compression from chest

Thyroid stimulation
Yogis claim that this pose stimulates and regulates your thyroid gland. However, unlike your digestive tract, the thyroid doesn't function through mechanical pressure and movement. Although it is possible that the increased blood flow could affect its function, scientific evidence does not seem to support these claims.

In this version, feet are pointed (in plantar flexion)

Hamstrings may engage slightly to maintain balance

Gluteal muscles may also engage minimally

Arms support your lower back

POSTERIOR–LATERAL VIEW

135

BRIDGE
Setu Bandhasana

Bridge is a gentle and accessible backbend that can help relieve back pain, particularly discomfort caused by sitting down too much. It is a calming pose, used by many to wind down at the end of a practice or at the end of the day in preparation for sleep.

THE BIG PICTURE

Bridge pose stretches the muscles along the front of your body – including your thighs, hips, abdomen, and chest. The back of your body strengthens – including your thighs, buttocks, back, and shoulders – as the muscles here work to support and hold you in an elevated backbend.

Torso
Your **spinal extensors** engage while your **abdominals** stretch. Your **pectorals** – particularly your **pectoralis minor muscles** – stretch as you broaden your chest. Your **middle** and **lower trapezius** work with the **rhomboids** to retract and stabilize your scapulae, while your **serratus anterior muscles** stretch.

Neck and arms
Your **cervical flexors** engage to flex your neck while your **cervical extensors** slightly stretch. Your **posterior deltoids**, **latissimus dorsi**, and **teres major muscles** engage to extend your shoulders. Your **triceps** extend your elbows.

KEY

•-- *Joints*

o-- *Muscles*

● Engaging

● Engaging while stretching

● Stretching

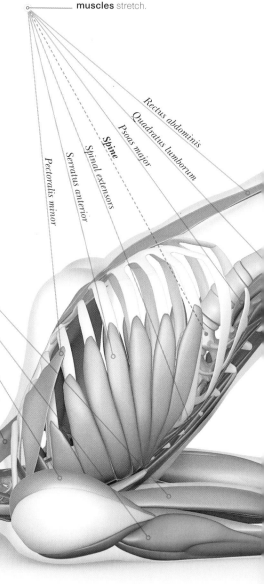

Rectus abdominis
Quadratus lumborum
Psoas major
Spine
Spinal extensors
Serratus anterior
Pectoralis minor

Biceps brachii
Triceps brachii
Deltoids
Sternocleidomastoid
Longus muscles
Splenius muscles

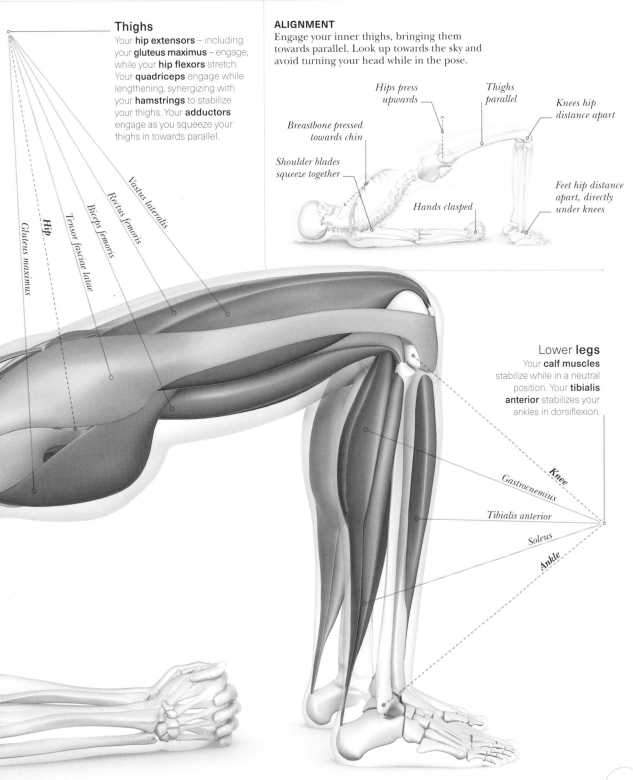

Thighs

Your **hip extensors** – including your **gluteus maximus** – engage, while your **hip flexors** stretch. Your **quadriceps** engage while lengthening, synergizing with your **hamstrings** to stabilize your thighs. Your **adductors** engage as you squeeze your thighs in towards parallel.

ALIGNMENT

Engage your inner thighs, bringing them towards parallel. Look up towards the sky and avoid turning your head while in the pose.

Hips press upwards

Thighs parallel

Knees hip distance apart

Breastbone pressed towards chin

Shoulder blades squeeze together

Hands clasped

Feet hip distance apart, directly under knees

Vastus lateralis

Rectus femoris

Biceps femoris

Tensor fasciae latae

Hip

Gluteus maximus

Lower legs

Your **calf muscles** stabilize while in a neutral position. Your **tibialis anterior** stabilizes your ankles in dorsiflexion.

Knee

Gastrocnemius

Tibialis anterior

Soleus

Ankle

137

» CLOSER LOOK

Backbends like this could also be considered "heart openers", because broadening your chest area may leave you feeling open-hearted. Your glutes strengthen and tone.

VARIATION

To challenge your pelvic stability, try raising one leg while in Bridge pose. Engage your core muscles to support your back as you lift one leg upwards. Focus on keeping your hips parallel. Press your standing foot into the floor to find support.

Leg lifts energetically upwards

Hips are extended

Rectus abdominis stretches

Send knees away from torso to find length through the hips

Hips press directly upwards

Heart pumps oxygenated blood around body

Lungs fill chest cavity as you inhale

Feel your ribcage expand three-dimensionally

Oxygenated blood flows to brain

Blood rush

Some claim that inversions send a rush of oxygenated blood to your head. This may happen briefly but your brain regulates its blood flow (see p.134). If you do get a head rush come down. Backbends are called "heart openers" because they create space in your chest. Be aware of this sensation and feel your ribcage expand as you inhale.

ANTERIOR VIEW

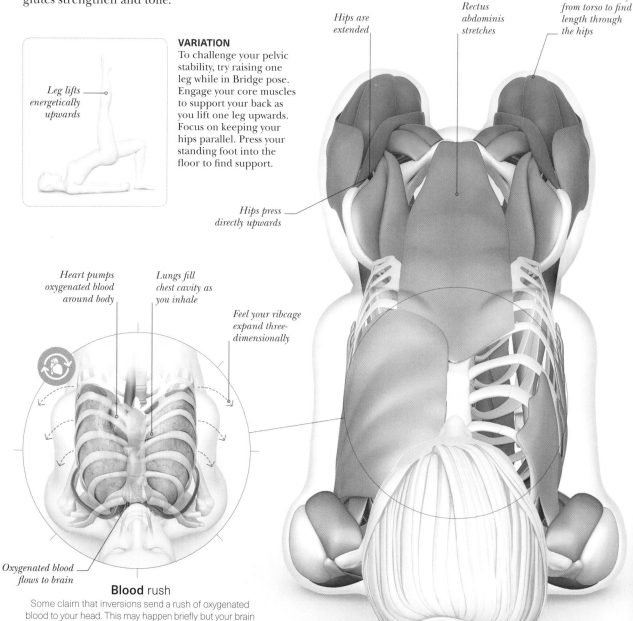

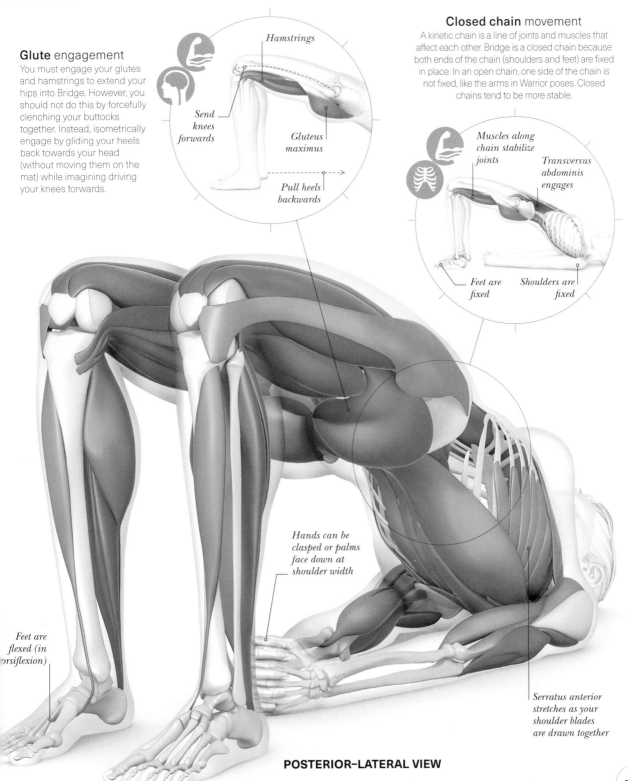

Glute engagement

You must engage your glutes and hamstrings to extend your hips into Bridge. However, you should not do this by forcefully clenching your buttocks together. Instead, isometrically engage by gliding your heels back towards your head (without moving them on the mat) while imagining driving your knees forwards.

Hamstrings

Send knees forwards

Gluteus maximus

Pull heels backwards

Closed chain movement

A kinetic chain is a line of joints and muscles that affect each other. Bridge is a closed chain because both ends of the chain (shoulders and feet) are fixed in place. In an open chain, one side of the chain is not fixed, like the arms in Warrior poses. Closed chains tend to be more stable.

Muscles along chain stabilize joints

Transversus abdominis engages

Feet are fixed

Shoulders are fixed

Hands can be clasped or palms face down at shoulder width

Feet are flexed (in orsiflexion)

Serratus anterior stretches as your shoulder blades are drawn together

POSTERIOR–LATERAL VIEW

WHEEL
Urdhva Dhanurasana

Wheel is a full backbend and inversion, bringing your head below the level of your heart. This pose is often done towards the end of a class as it requires warming up to be safe for most people. With practice, Wheel can improve the strength and flexibility of your back.

THE BIG PICTURE

This pose strongly stretches the muscles at the front of your body – including your thighs, hips, abdomen, and chest. It strengthens your shoulders and the back of your body – particularly your back muscles, buttocks, and thighs – as they support you in this deep backbend and elevation.

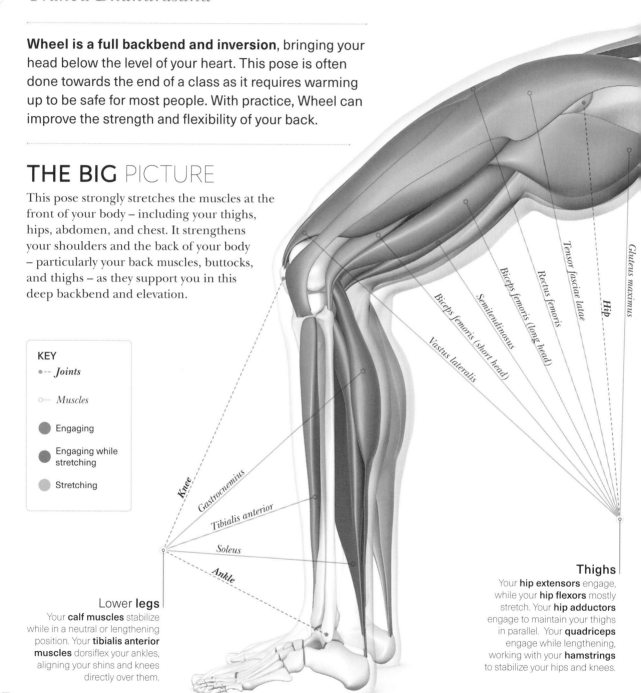

KEY

●--- *Joints*

○--- *Muscles*

● Engaging

● Engaging while stretching

● Stretching

Gluteus maximus

Hip

Tensor fasciae latae

Rectus femoris

Biceps femoris (long head)

Semitendinosus

Biceps femoris (short head)

Vastus lateralis

Knee

Gastrocnemius

Tibialis anterior

Soleus

Ankle

Lower **legs**
Your **calf muscles** stabilize while in a neutral or lengthening position. Your **tibialis anterior muscles** dorsiflex your ankles, aligning your shins and knees directly over them.

Thighs
Your **hip extensors** engage, while your **hip flexors** mostly stretch. Your **hip adductors** engage to maintain your thighs in parallel. Your **quadriceps** engage while lengthening, working with your **hamstrings** to stabilize your hips and knees.

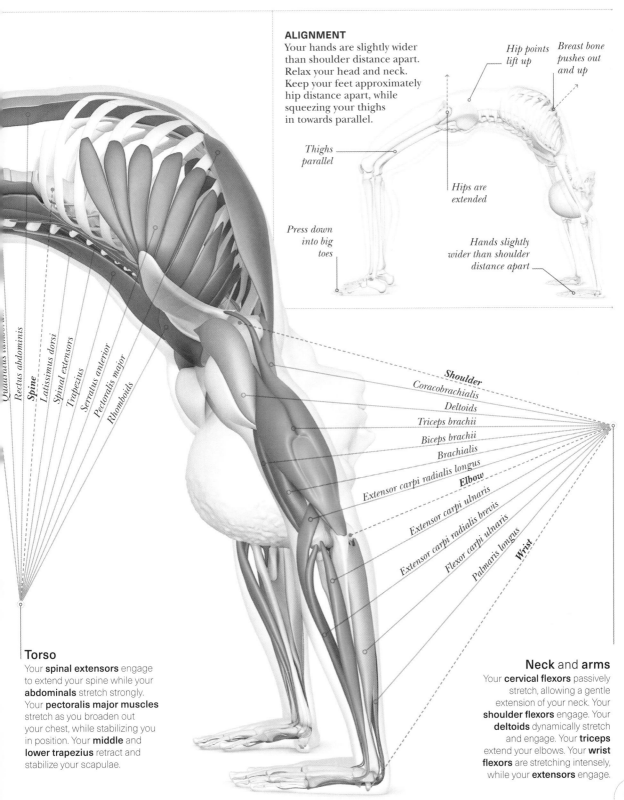

Your hands are slightly wider
than shoulder distance apart.
Relax your head and neck.
Keep your feet approximately
hip distance apart, while
squeezing your thighs
in towards parallel.

*Hip points
lift up*

*Breast bone
pushes out
and up*

*Thighs
parallel*

*Hips are
extended*

*Press down
into big
toes*

*Hands slightly
wider than shoulder
distance apart*

Quadratus lumbor...

Rectus abdominis

Spine

Latissimus dorsi

Spinal extensors

Trapezius

Serratus anterior

Pectoralis major

Rhomboids

Shoulder

Coracobrachialis

Deltoids

Triceps brachii

Biceps brachii

Brachialis

Extensor carpi radialis longus

Elbow

Extensor carpi ulnaris

Extensor carpi radialis brevis

Flexor carpi ulnaris

Palmaris longus

Wrist

Torso

Your **spinal extensors** engage
to extend your spine while your
abdominals stretch strongly.
Your **pectoralis major muscles**
stretch as you broaden out
your chest, while stabilizing you
in position. Your **middle** and
lower trapezius retract and
stabilize your scapulae.

Neck and arms

Your **cervical flexors** passively
stretch, allowing a gentle
extension of your neck. Your
shoulder flexors engage. Your
deltoids dynamically stretch
and engage. Your **triceps**
extend your elbows. Your **wrist
flexors** are stretching intensely,
while your **extensors** engage.

141

›› CLOSER LOOK

Wheel puts your shoulder joints and spine in a unique position that can be challenging both for people who are tight and those who are very flexible. It can be quite demanding, yet energizing and uplifting.

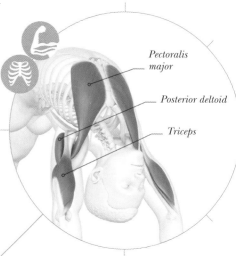

Pectoralis major

Posterior deltoid

Triceps

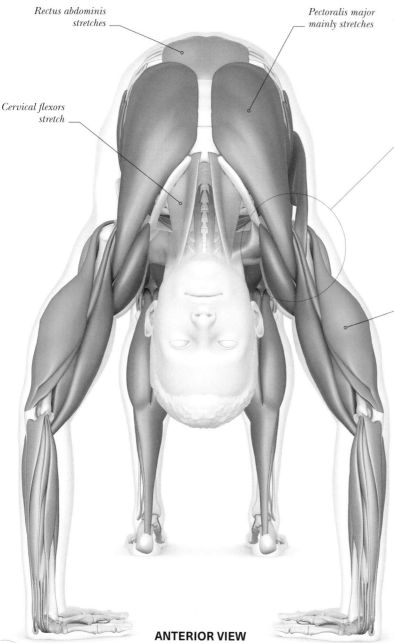

Rectus abdominis stretches

Pectoralis major mainly stretches

Cervical flexors stretch

ANTERIOR VIEW

Tight **shoulders**

Tight shoulders are a common limiting factor for this pose. Many people lack the range of motion to go into full shoulder flexion, bringing the arms directly overhead. Make sure you thoroughly warm up your shoulders before doing Wheel. Stretch your shoulders over time with poses like Cow Face (see pp.60–63).

Triceps engage to extend elbows, but may stretch too if tight

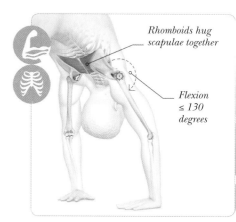

Rhomboids hug scapulae together

Flexion ≤ 130 degrees

Shoulder flexion

In shoulder flexion there is little stability, particularly when weight bearing as in Wheel. If you are very flexible, especially with a tendency towards dislocation, be mindful when doing this pose, or try Bridge pose for a more stable shoulder position in extension instead (see pp.136–39).

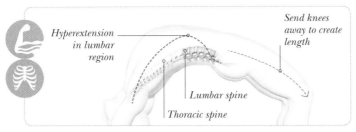

Hyperextension in lumbar region

Send knees away to create length

Lumbar spine

Thoracic spine

Spinal flexibility

Many yogis allow too much bending or hyperextension in the lower back, like this. If this is you, focus on lengthening your lower back instead of crunching and sinking into it. Although your lumbar spine has a greater capacity for extension than your thoracic, try to make the extension more even.

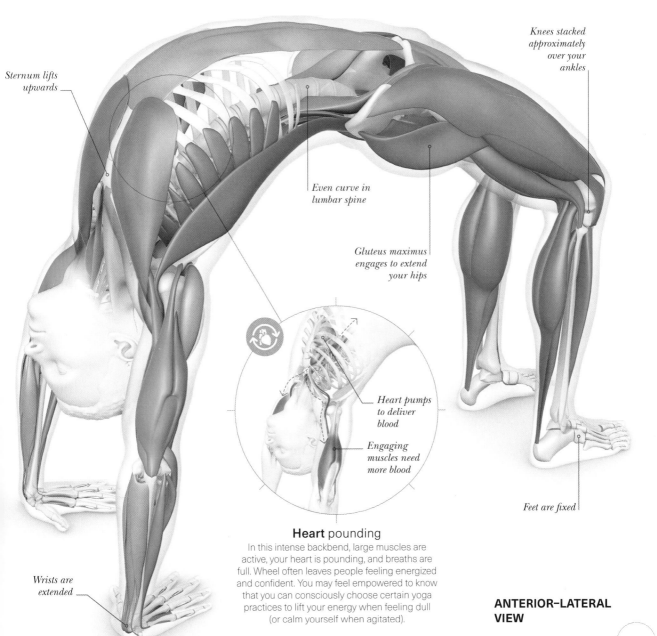

Sternum lifts upwards

Even curve in lumbar spine

Gluteus maximus engages to extend your hips

Knees stacked approximately over your ankles

Heart pumps to deliver blood

Engaging muscles need more blood

Wrists are extended

Feet are fixed

Heart pounding

In this intense backbend, large muscles are active, your heart is pounding, and breaths are full. Wheel often leaves people feeling energized and confident. You may feel empowered to know that you can consciously choose certain yoga practices to lift your energy when feeling dull (or calm yourself when agitated).

ANTERIOR–LATERAL VIEW

143

FLOOR ASANAS

These floor asanas include arm balance poses, prone (on your front) poses, and supine (on your back) poses. The asanas in this section range from intense and strong, like Plank, to soft and subtle, like the Supine Twist. No matter the intensity, they all provide a rich opportunity to inquire within yourself.

CROW

Bakasana

Crow pose is an arm balance that uniquely develops your strength, flexibility, balance, and agility. Working your wrist muscles is a great antidote to typing on a computer all day. Plus this challenging pose offers an opportunity to face your fears and be playful in your practice.

THE BIG PICTURE

Practising Crow strengthens the muscles of your wrists, shoulders, arms, hips, and abdomen. In this pose, you are fully weight-bearing on your hands, with your upper body working to support you and keep you balanced.

KEY

●--- *Joints*

○--- *Muscles*

● Engaging

● Engaging while stretching

● Stretching

Thighs

Your **hip flexors** engage to flex your hips. Your **hamstrings** flex your knees and your **quadriceps** stretch. Your **adductors** are recruited to adduct and stabilize your hips and thighs.

Hip
Tensor fasciae latae
Semitendinosus
Biceps femoris
Rectus femoris
Vastus lateralis

Ankle
Flexor hallucis longus
Soleus
Gastrocnemius
Tibialis anterior
Knee

Lower **legs**

Your **plantar flexors** engage to point your toes, while your **dorsiflexors** slightly stretch – particularly your **tibialis anterior**.

ALIGNMENT

Your knees rest on a shelf created by your upper arms. Gaze forwards with your chin slightly lifted. Press down into the floor and be prepared to fall backwards with grace.

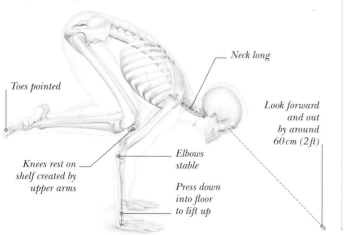

Toes pointed

Neck long

Look forward and out by around 60 cm (2 ft)

Knees rest on shelf created by upper arms

Elbows stable

Press down into floor to lift up

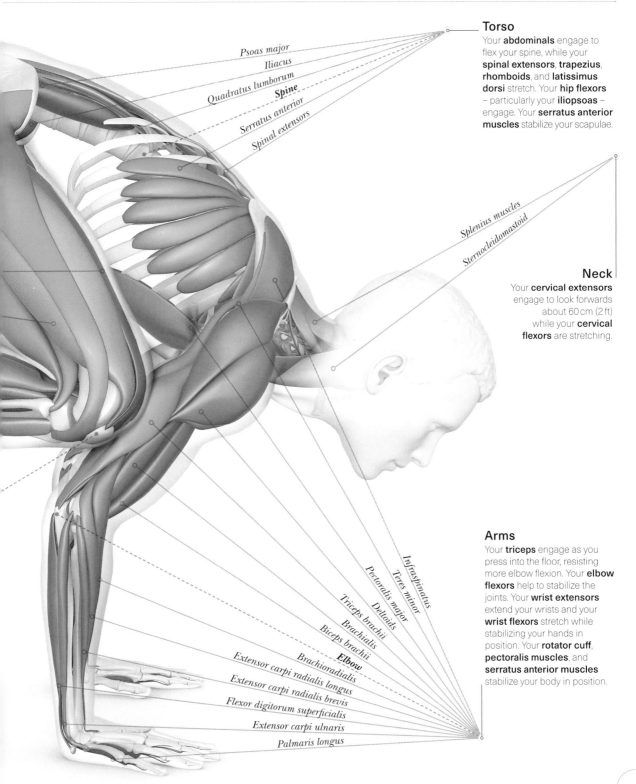

Psoas major
Iliacus
Quadratus lumborum
Spine
Serratus anterior
Spinal extensors

Splenius muscles
Sternocleidomastoid

Infraspinatus
Teres minor
Pectoralis major
Triceps brachii
Deltoids
Brachialis
Biceps brachii
Elbow
Brachioradialis
Extensor carpi radialis longus
Extensor carpi radialis brevis
Flexor digitorum superficialis
Extensor carpi ulnaris
Palmaris longus

Torso

Your **abdominals** engage to flex your spine, while your **spinal extensors, trapezius, rhomboids,** and **latissimus dorsi** stretch. Your **hip flexors** – particularly your **iliopsoas** – engage. Your **serratus anterior muscles** stabilize your scapulae.

Neck

Your **cervical extensors** engage to look forwards about 60 cm (2 ft) while your **cervical flexors** are stretching.

Arms

Your **triceps** engage as you press into the floor, resisting more elbow flexion. Your **elbow flexors** help to stabilize the joints. Your **wrist extensors** extend your wrists and your **wrist flexors** stretch while stabilizing your hands in position. Your **rotator cuff, pectoralis muscles,** and **serratus anterior muscles** stabilize your body in position.

» **CLOSER** LOOK

Crow is a challenging balancing pose that strengthens
your wrists. Finding playfulness in the pose can help you
reveal a sense of bravery and resilience.

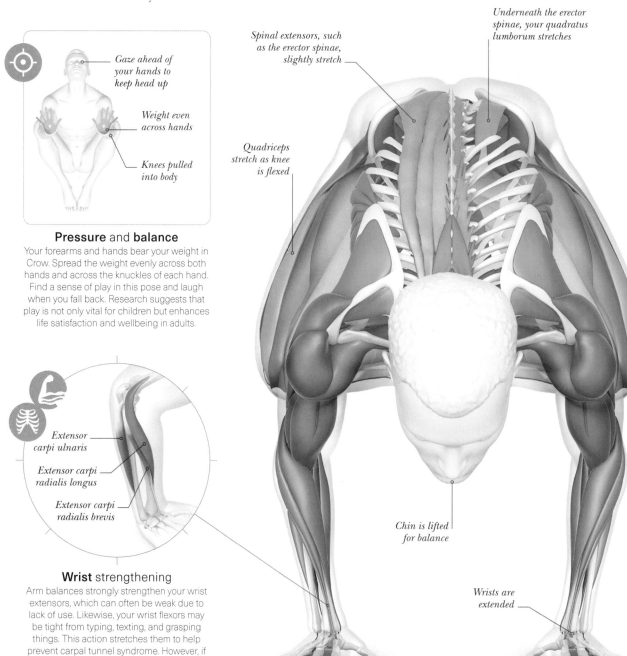

Gaze ahead of
your hands to
keep head up

Weight even
across hands

Knees pulled
into body

Spinal extensors, such
as the erector spinae,
slightly stretch

Underneath the erector
spinae, your quadratus
lumborum stretches

Quadriceps
stretch as knee
is flexed

Chin is lifted
for balance

Wrists are
extended

Extensor
carpi ulnaris

Extensor carpi
radialis longus

Extensor carpi
radialis brevis

ANTERIOR VIEW

Pressure and **balance**

Your forearms and hands bear your weight in
Crow. Spread the weight evenly across both
hands and across the knuckles of each hand.
Find a sense of play in this pose and laugh
when you fall back. Research suggests that
play is not only vital for children but enhances
life satisfaction and wellbeing in adults.

Wrist strengthening

Arm balances strongly strengthen your wrist
extensors, which can often be weak due to
lack of use. Likewise, your wrist flexors may
be tight from typing, texting, and grasping
things. This action stretches them to help
prevent carpal tunnel syndrome. However, if
you currently have wrist issues, this amount
of weight bearing is probably too much.

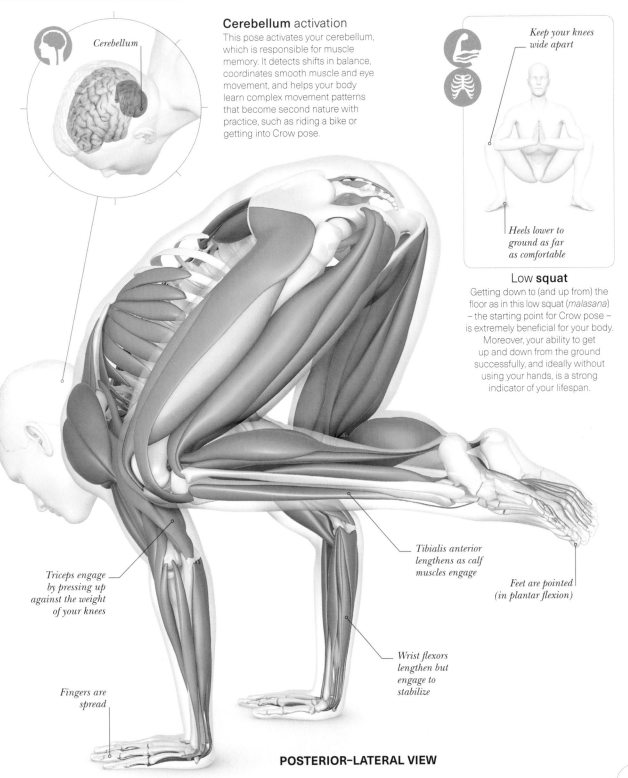

Cerebellum activation

This pose activates your cerebellum, which is responsible for muscle memory. It detects shifts in balance, coordinates smooth muscle and eye movement, and helps your body learn complex movement patterns that become second nature with practice, such as riding a bike or getting into Crow pose.

Cerebellum

Keep your knees wide apart

Heels lower to ground as far as comfortable

Low **squat**

Getting down to (and up from) the floor as in this low squat (*malasana*) – the starting point for Crow pose – is extremely beneficial for your body. Moreover, your ability to get up and down from the ground successfully, and ideally without using your hands, is a strong indicator of your lifespan.

Triceps engage by pressing up against the weight of your knees

Fingers are spread

Tibialis anterior lengthens as calf muscles engage

Feet are pointed (in plantar flexion)

Wrist flexors lengthen but engage to stabilize

POSTERIOR-LATERAL VIEW

PLANK
Kumbhakasana

Plank pose is the top point of a press up. It is a strong, stabilizing pose that works muscles from the deepest layers inside you to the most superficial. When holding Plank you are giving your body a thorough, strengthening workout.

THE BIG PICTURE

Plank pose particularly strengthens your shoulders and entire core – including your abdominals, back muscles, and pelvic floor muscles. It builds heat and energy throughout your body when held for several breaths or more.

Thighs

Your **quadriceps** engage to extend your knees and stabilize your thighs. Your **hip adductors** and **abductors** engage while in a neutral position to stabilize your thighs and hips.

Gluteus medius
Gluteus maximus
Tensor fasciae latae
Rectus femoris
Semitendinosus
Biceps femoris
Vastus lateralis
Knee

Lower legs

Your **ankle dorsiflexors** engage, as you press your heels back. You are likely to feel a stretch in your **toe flexors** and in the plantar region of your foot. Your **calf muscles** are in a slightlly stretched position.

Gastrocnemius
Tibilias anterior
Extensor digitorum longus
Extensor hallucis longus
Soleus
Ankle
Plantar fascia

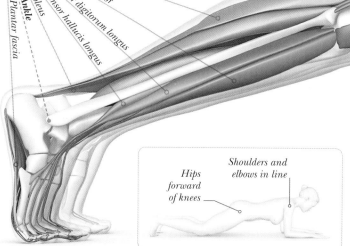

Hips forward of knees

Shoulders and elbows in line

VARIATION

Place your forearms and, optionally, knees down to lower the intensity. Don't allow your back to sag – if you feel any strain on your back come out of the pose and rest.

KEY

●-- *Joints*

○— *Muscles*

● Engaging

● Engaging while stretching

● Stretching

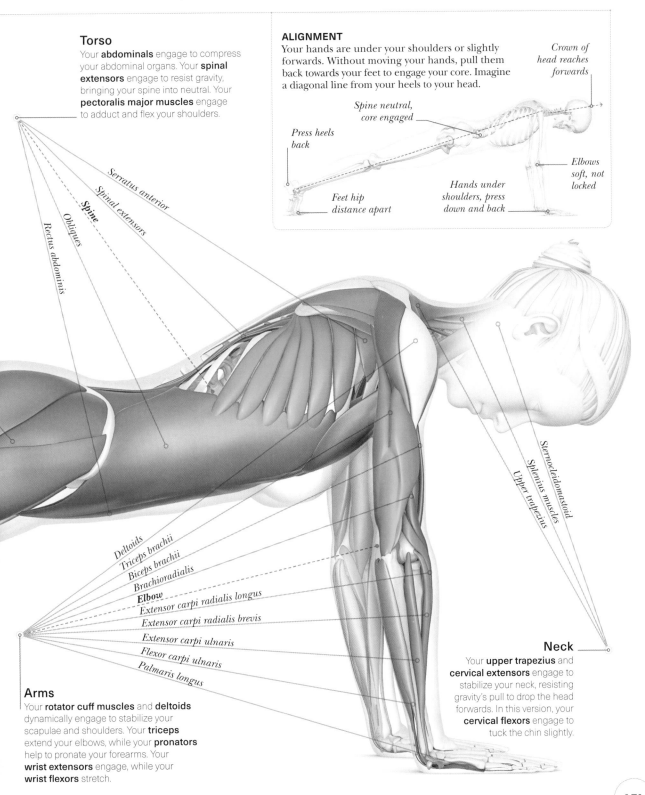

Torso

Your **abdominals** engage to compress your abdominal organs. Your **spinal extensors** engage to resist gravity, bringing your spine into neutral. Your **pectoralis major muscles** engage to adduct and flex your shoulders.

Serratus anterior

Spinal extensors

Spine

Obliques

Rectus abdominis

Sternocleidomastoid

Splenius muscles

Upper trapezius

Deltoids

Triceps brachii

Biceps brachii

Brachioradialis

Elbow

Extensor carpi radialis longus

Extensor carpi radialis brevis

Extensor carpi ulnaris

Flexor carpi ulnaris

Palmaris longus

Neck

Your **upper trapezius** and **cervical extensors** engage to stabilize your neck, resisting gravity's pull to drop the head forwards. In this version, your **cervical flexors** engage to tuck the chin slightly.

Arms

Your **rotator cuff muscles** and **deltoids** dynamically engage to stabilize your scapulae and shoulders. Your **triceps** extend your elbows, while your **pronators** help to pronate your forearms. Your **wrist extensors** engage, while your **wrist flexors** stretch.

151

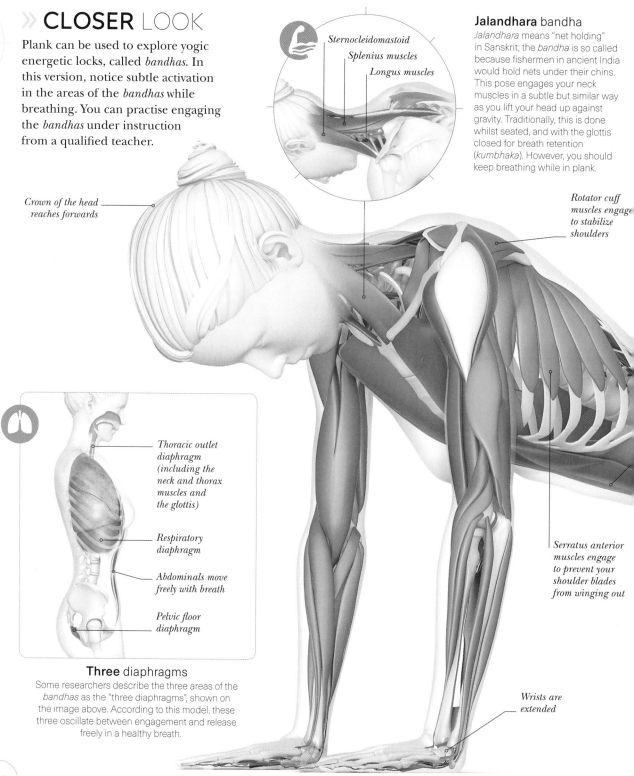

» CLOSER LOOK

Plank can be used to explore yogic energetic locks, called *bandhas*. In this version, notice subtle activation in the areas of the *bandhas* while breathing. You can practise engaging the *bandhas* under instruction from a qualified teacher.

Sternocleidomastoid

Splenius muscles

Longus muscles

Jalandhara bandha

Jalandhara means "net holding" in Sanskrit; the *bandha* is so called because fishermen in ancient India would hold nets under their chins. This pose engages your neck muscles in a subtle but similar way as you lift your head up against gravity. Traditionally, this is done whilst seated, and with the glottis closed for breath retention (*kumbhaka*). However, you should keep breathing while in plank.

Crown of the head reaches forwards

Rotator cuff muscles engage to stabilize shoulders

Thoracic outlet diaphragm (including the neck and thorax muscles and the glottis)

Respiratory diaphragm

Abdominals move freely with breath

Pelvic floor diaphragm

Serratus anterior muscles engage to prevent your shoulder blades from winging out

Three diaphragms

Some researchers describe the three areas of the *bandhas* as the "three diaphragms", shown on the image above. According to this model, these three oscillate between engagement and release freely in a healthy breath.

Wrists are extended

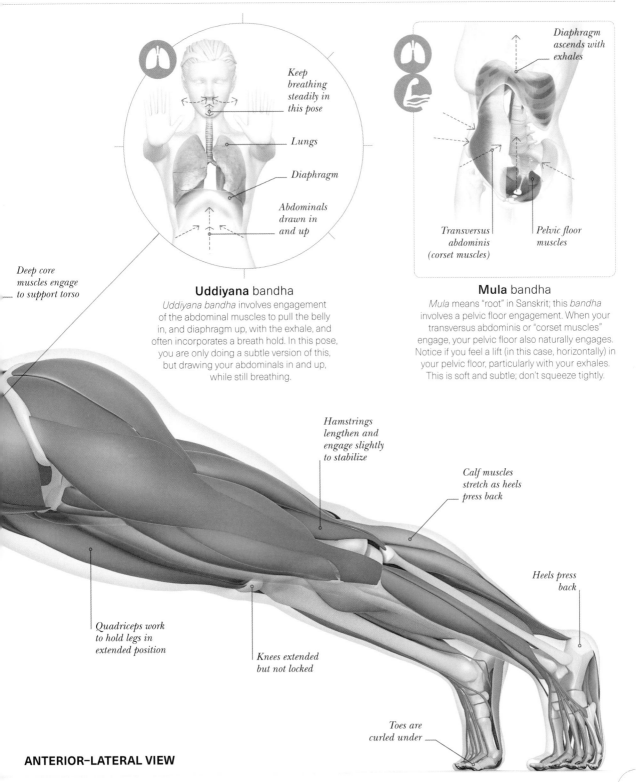

Keep breathing steadily in this pose

Lungs

Diaphragm

Abdominals drawn in and up

Diaphragm ascends with exhales

Transversus abdominis (corset muscles)

Pelvic floor muscles

Uddiyana bandha

Uddiyana bandha involves engagement of the abdominal muscles to pull the belly in, and diaphragm up, with the exhale, and often incorporates a breath hold. In this pose, you are only doing a subtle version of this, but drawing your abdominals in and up, while still breathing.

Mula bandha

Mula means "root" in Sanskrit; this *bandha* involves a pelvic floor engagement. When your transversus abdominis or "corset muscles" engage, your pelvic floor also naturally engages. Notice if you feel a lift (in this case, horizontally) in your pelvic floor, particularly with your exhales. This is soft and subtle; don't squeeze tightly.

Deep core muscles engage to support torso

Hamstrings lengthen and engage slightly to stabilize

Calf muscles stretch as heels press back

Heels press back

Quadriceps work to hold legs in extended position

Knees extended but not locked

Toes are curled under

ANTERIOR-LATERAL VIEW

153

SIDE PLANK
Vasisthasana

Side plank is a challenging arm balance that may get you sweating and your heart pounding. This pose is particularly beneficial for anyone looking to improve their focus and endurance. Holding Side plank takes concentration to avoid sagging your hips down.

THE BIG PICTURE

This pose strengthens your core – including your abdominals and back muscles. Your supporting arm and shoulder muscles are also engaging strongly to maintain balance. Even your leg muscles are working to support you and keep you aligned and balanced.

KEY
- --- *Joints*
- ○— *Muscles*
- ⬤ Engaging
- ⬤ Engaging while stretching
- ⬤ Stretching

Top **thigh**
Your **hip adductors** engage on both sides to stabilize your thighs.

Lower **legs**
Your **ankle dorsiflexors** engage to dorsiflex your ankles and extend your toes. Your **calf muscles** are in a stretched position. Press the side of your foot into the floor to activate your **fibularis muscles** in your bottom leg, preventing your ankle from rolling downwards.

ALIGNMENT
Try to stack your hips and shoulders on top of each other. If comfortable, reach your top arm up and gaze skyward. Alternatively, you may find looking down at your supporting hand helps you to stay balanced.

Hand reaches up

Gaze up

Shoulders and hips stacked

Hips lift up

Feet stacked

Elbows soft, not locked

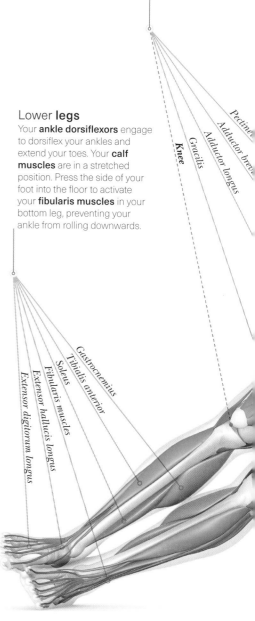

Pectine...
Adductor brev...
Adductor longus
Gracilis
Knee
Gastrocnemius
Tibialis anterior
Soleus
Fibularis muscles
Extensor hallucis longus
Extensor digitorum longus

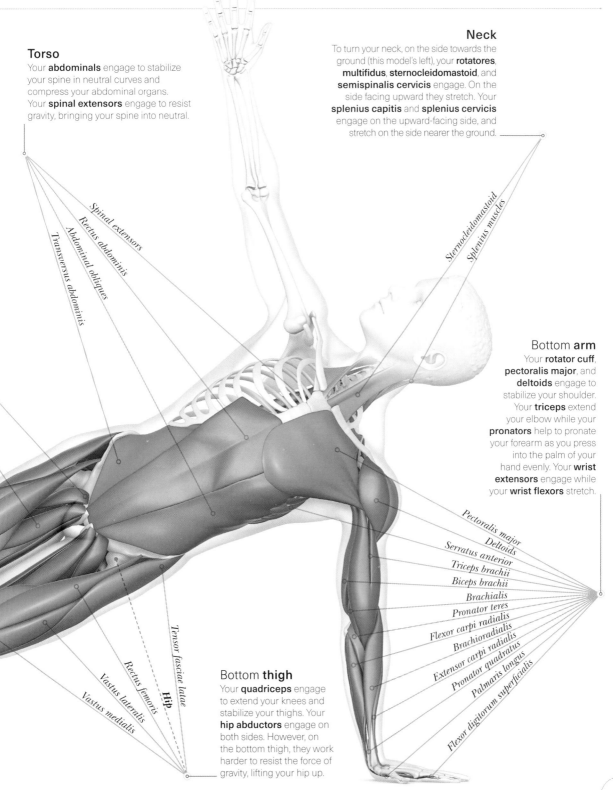

Torso

Your **abdominals** engage to stabilize your spine in neutral curves and compress your abdominal organs. Your **spinal extensors** engage to resist gravity, bringing your spine into neutral.

Spinal extensors
Rectus abdominis
Abdominal obliques
Transversus abdominis

Neck

To turn your neck, on the side towards the ground (this model's left), your **rotatores**, **multifidus**, **sternocleidomastoid**, and **semispinalis cervicis** engage. On the side facing upward they stretch. Your **splenius capitis** and **splenius cervicis** engage on the upward-facing side, and stretch on the side nearer the ground.

Sternocleidomastoid
Splenius muscles

Bottom **arm**

Your **rotator cuff**, **pectoralis major**, and **deltoids** engage to stabilize your shoulder. Your **triceps** extend your elbow while your **pronators** help to pronate your forearm as you press into the palm of your hand evenly. Your **wrist extensors** engage while your **wrist flexors** stretch.

Pectoralis major
Deltoids
Serratus anterior
Triceps brachii
Biceps brachii
Brachialis
Pronator teres
Flexor carpi radialis
Brachioradialis
Extensor carpi radialis
Pronator quadratus
Palmaris longus
Flexor digitorum superficialis

Bottom **thigh**

Your **quadriceps** engage to extend your knees and stabilize your thighs. Your **hip abductors** engage on both sides. However, on the bottom thigh, they work harder to resist the force of gravity, lifting your hip up.

Tensor fasciae latae
Hip
Rectus femoris
Vastus lateralis
Vastus medialis

155

»CLOSER LOOK

Side Plank involves deep breaths, recruiting
more respiratory muscles than usual. There is
also significant core muscle engagement, which
is good for scoliosis, but poses risks for pregnancy.

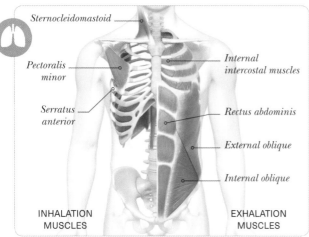

Sternocleidomastoid

Pectoralis
minor

Serratus
anterior

Internal
intercostal muscles

Rectus abdominis

External oblique

Internal oblique

INHALATION
MUSCLES

EXHALATION
MUSCLES

Respiratory muscles

In a natural breath, your diaphragm is the main player.
When you breathe deeply, as in this pose, other accessory
respiratory muscles can be recruited. The inhale involves
the muscles above left, along with small muscles along
your neck called the scalenes. The exhale also involves
deep muscles along your ribs called transversus thoracis.

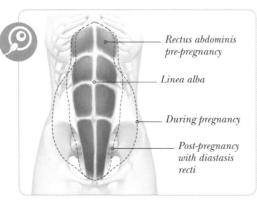

Rectus abdominis
pre-pregnancy

Linea alba

During pregnancy

Post-pregnancy
with diastasis
recti

Pregnancy caution

The linea alba is the connective tissue joining the sections
of the rectus abdominis. During pregnancy, pressure can
separate this tissue causing a condition called diastasis
recti or abdominal separation. For this reason, be cautious
of poses that involve abdominal engagement and
pressure while pregnant, particularly later in pregnancy.

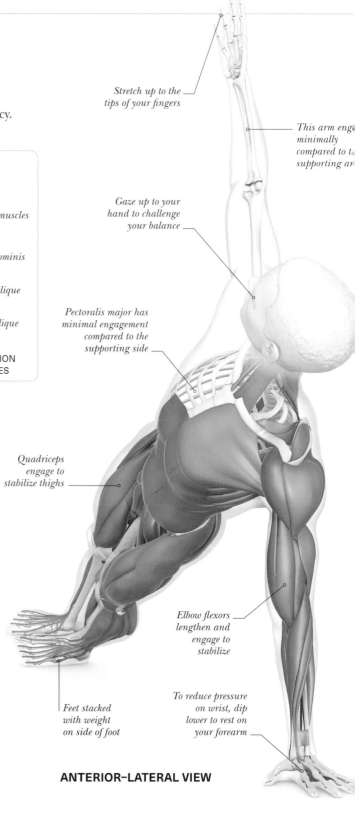

Stretch up to the
tips of your fingers

This arm eng
minimally
compared to t
supporting ar

Gaze up to your
hand to challenge
your balance

Pectoralis major has
minimal engagement
compared to the
supporting side

Quadriceps
engage to
stabilize thighs

Elbow flexors
lengthen and
engage to
stabilize

Feet stacked
with weight
on side of foot

To reduce pressure
on wrist, dip
lower to rest on
your forearm

ANTERIOR–LATERAL VIEW

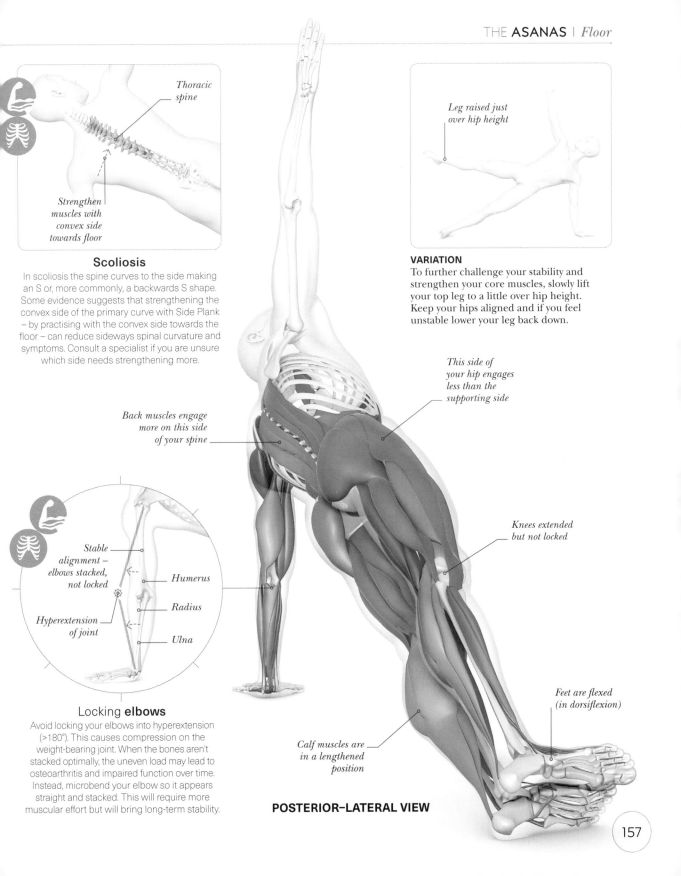

Thoracic spine

Scoliosis

Strengthen muscles with convex side towards floor

In scoliosis the spine curves to the side making an S or, more commonly, a backwards S shape. Some evidence suggests that strengthening the convex side of the primary curve with Side Plank – by practising with the convex side towards the floor – can reduce sideways spinal curvature and symptoms. Consult a specialist if you are unsure which side needs strengthening more.

Leg raised just over hip height

VARIATION
To further challenge your stability and strengthen your core muscles, slowly lift your top leg to a little over hip height. Keep your hips aligned and if you feel unstable lower your leg back down.

This side of your hip engages less than the supporting side

Back muscles engage more on this side of your spine

Knees extended but not locked

Stable alignment – elbows stacked, not locked

Humerus

Radius

Hyperextension of joint

Ulna

Feet are flexed (in dorsiflexion)

Locking **elbows**

Avoid locking your elbows into hyperextension (>180°). This causes compression on the weight-bearing joint. When the bones aren't stacked optimally, the uneven load may lead to osteoarthritis and impaired function over time. Instead, microbend your elbow so it appears straight and stacked. This will require more muscular effort but will bring long-term stability.

Calf muscles are in a lengthened position

POSTERIOR-LATERAL VIEW

157

COBRA
Bhujangasana

Cobra pose is a key traditional yoga pose. This gentle backbend was believed to ignite a burning digestive fire and awaken the flow of dormant energy. It does seem to stimulate digestion and elimination, while helping to ease back pain for many.

THE BIG PICTURE

The front of your body – including your chest, abdominals, and hips – is stretching. Meanwhile, muscles in your back, shoulders, and arms are strengthening as you maintain the posture, creating an even curve along your neck and spine.

Neck long and chin lifted

VARIATION
With your forearms on the floor, Sphinx can be a more accessible and more passive version of the pose.

Elbows un shoulders

Thighs
Your **gluteus maximus**, **adductor magnus**, and **hamstrings** engage to hold your hips in extension, while your **tensor fasciae latae** and **iliotibial band** stabilize your hips.

ALIGNMENT
Your pubic bone remains on the mat as you elongate your spine into an even backbend. If you feel pinching or pain in your lower back, come down lower.

Gaze straight ahead

Shoulder blades down and towards centre

Even curve in spine

No pinching in lower back

Toes untucked

Buttocks not clenched

Even curve in neck

Breastbone reaches forwards and up

Elbows bent

Pubis remains on mat

Gluteus maximus
Hip
Tensor fasciae latae
Biceps femoris
Semitendinosus
Vastus lateralis
Iliotibial band

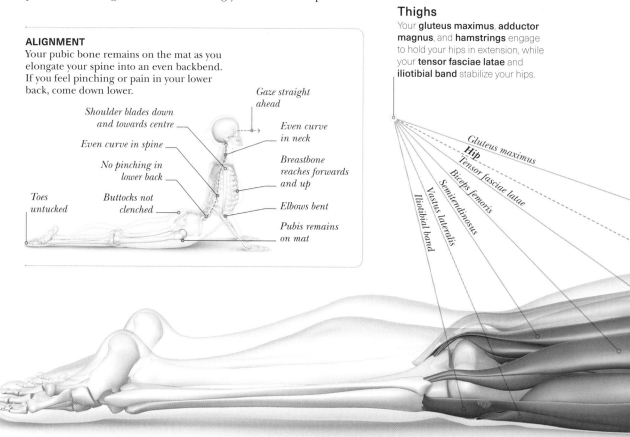

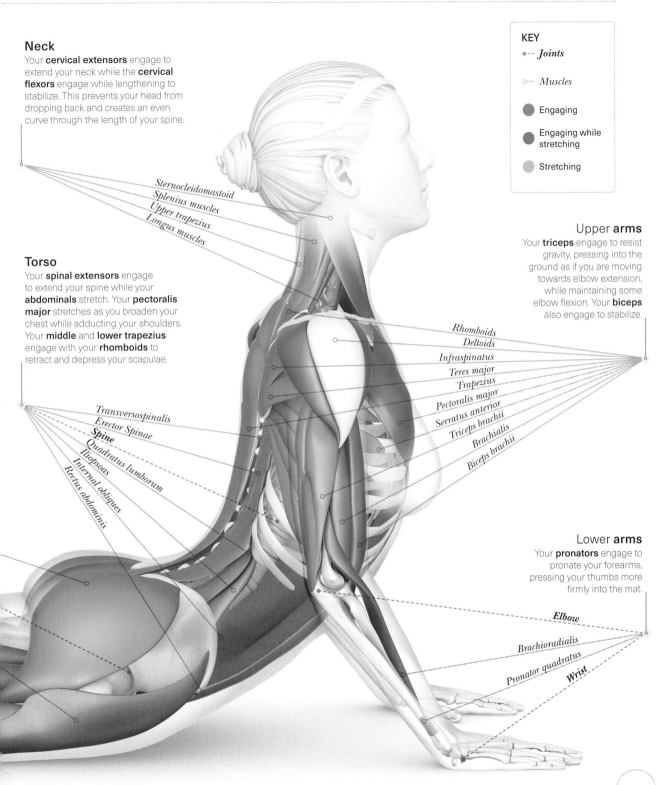

Neck

Your **cervical extensors** engage to extend your neck while the **cervical flexors** engage while lengthening to stabilize. This prevents your head from dropping back and creates an even curve through the length of your spine.

Torso

Your **spinal extensors** engage to extend your spine while your **abdominals** stretch. Your **pectoralis major** stretches as you broaden your chest while adducting your shoulders. Your **middle** and **lower trapezius** engage with your **rhomboids** to retract and depress your scapulae.

Upper **arms**

Your **triceps** engage to resist gravity, pressing into the ground as if you are moving towards elbow extension, while maintaining some elbow flexion. Your **biceps** also engage to stabilize.

Lower **arms**

Your **pronators** engage to pronate your forearms, pressing your thumbs more firmly into the mat.

Sternocleidomastoid
Splenius muscles
Upper trapezius
Longus muscles

Transversospinalis
Erector Spinae
Spine
Quadratus lumborum
Iliopsoas
Internal obliques
Rectus abdominis

Rhomboids
Deltoids
Infraspinatus
Teres major
Trapezius
Pectoralis major
Serratus anterior
Triceps brachii
Brachialis
Biceps brachii

Elbow
Brachioradialis
Pronator quadratus
Wrist

≫ CLOSER LOOK

Cobra pose can be refined with activation of key muscles like the serratus anterior. It can be adapted to gentler versions such as Sphinx pose, or a deeper backbend such as Upward-facing Dog.

Keep spine gently curved to protect discs

Blood vessels close to spine can be damaged in hyperextension

Hyperextension could increase fluid pressure in eyes

Eyes gaze up diagonally to where the ceiling meets the wall

Chin lifts slightly

Pectoralis minor mainly stretches

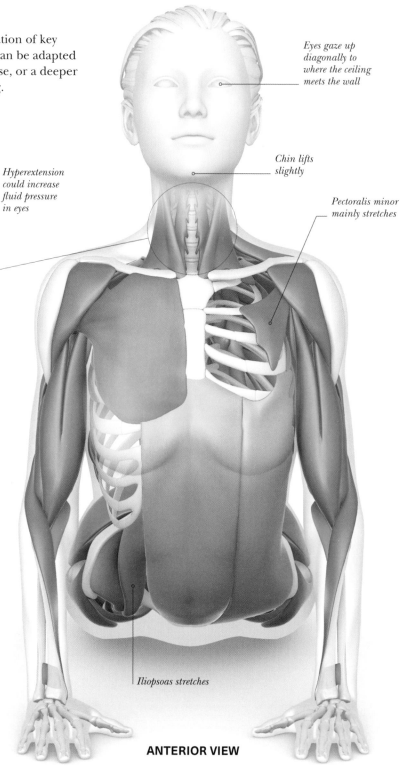

ANTERIOR VIEW

Iliopsoas stretches

Neck problems

Traditional teachers have taught to throw the head back as far as possible. However, we now understand this has more risks than benefits. Based on case studies and anatomy knowledge, you can choose to respectfully adapt for safety and optimal function by avoiding hyperextension.

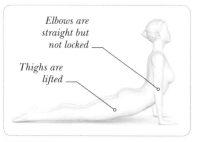

Elbows are straight but not locked

Thighs are lifted

VARIATION

Upward-facing Dog is a similar pose to Cobra, used more in some styles of yoga. The thighs are lifted off the floor with the elbows straight to create a deeper backbend.

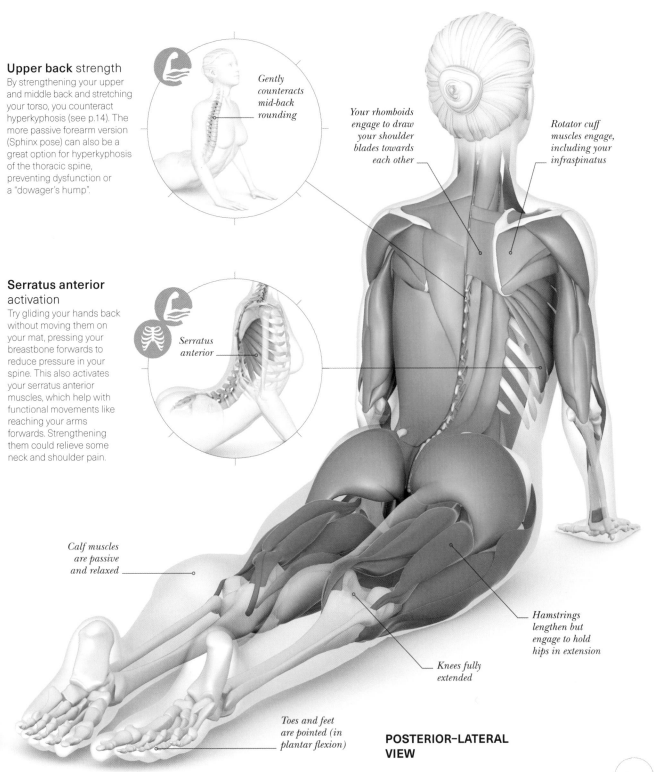

Upper back strength

By strengthening your upper and middle back and stretching your torso, you counteract hyperkyphosis (see p.14). The more passive forearm version (Sphinx pose) can also be a great option for hyperkyphosis of the thoracic spine, preventing dysfunction or a "dowager's hump".

Gently counteracts mid-back rounding

Serratus anterior activation

Try gliding your hands back without moving them on your mat, pressing your breastbone forwards to reduce pressure in your spine. This also activates your serratus anterior muscles, which help with functional movements like reaching your arms forwards. Strengthening them could relieve some neck and shoulder pain.

Serratus anterior

Your rhomboids engage to draw your shoulder blades towards each other

Rotator cuff muscles engage, including your infraspinatus

Calf muscles are passive and relaxed

Hamstrings lengthen but engage to hold hips in extension

Knees fully extended

Toes and feet are pointed (in plantar flexion)

POSTERIOR–LATERAL VIEW

161

LOCUST
Salabhasana

Locust pose, also known as belly-down boat pose, can be helpful for relieving back pain. Elongating your spine in this way helps to counteract poor posture and related issues, as muscles along your back and legs engage to hold each end of your body off the ground.

THE BIG PICTURE

This pose particularly strengthens your back muscles, buttocks, and thighs as you lift your legs and shoulders from the ground. It can be challenging, but you don't have to lift very high to get the benefits of the pose.

Thighs
Your **hip extensors** engage to help you lift your thighs, while your **hip flexors** stretch. Your **quadriceps** engage to extend your knees.

Iliopsoas

Gluteus maximus

Hip

Tensor fasciae latae

Rectus femoris

Vastus lateralis

Biceps femoris (long head)

Biceps femoris (short head)

Ankle

Soleus

Gastrocnemius

Tibialis anterior

Knee

Lower legs
Your **gastrocnemius** and **soleus** engage to plantarflex your ankles while your **tibialis anterior muscles** and other **ankle dorsiflexors** are in a stretched position.

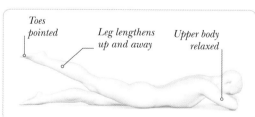

Toes pointed

Leg lengthens up and away

Upper body relaxed

VARIATION
If you have neck issues, place your forehead on your hands and lift one leg at a time, trying to keep both of your front hip points towards the ground. Hold for several breaths, then switch legs.

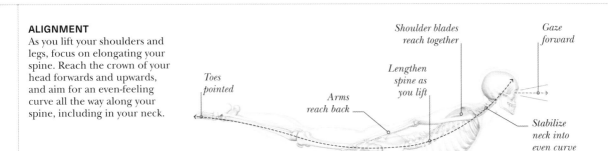

ALIGNMENT
As you lift your shoulders and legs, focus on elongating your spine. Reach the crown of your head forwards and upwards, and aim for an even-feeling curve all the way along your spine, including in your neck.

Toes pointed

Arms reach back

Shoulder blades reach together

Lengthen spine as you lift

Gaze forward

Stabilize neck into even curve

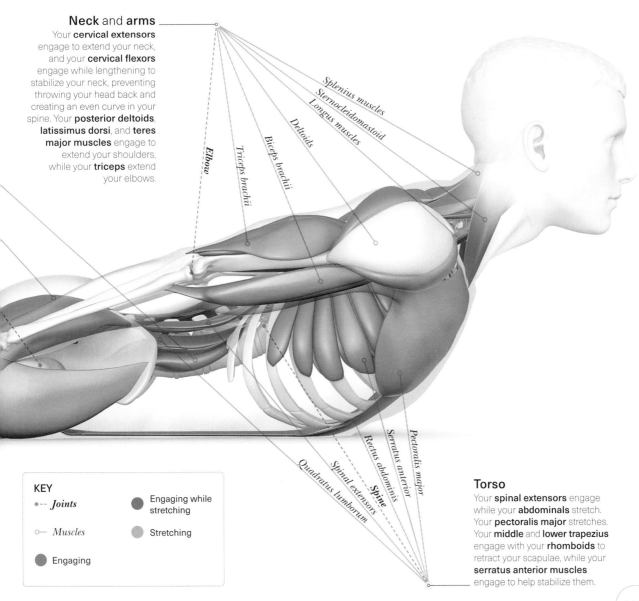

Neck and arms
Your **cervical extensors** engage to extend your neck, and your **cervical flexors** engage while lengthening to stabilize your neck, preventing throwing your head back and creating an even curve in your spine. Your **posterior deltoids, latissimus dorsi**, and **teres major muscles** engage to extend your shoulders, while your **triceps** extend your elbows.

Splenius muscles
Sternocleidomastoid
Longus muscles
Deltoids
Biceps brachii
Triceps brachii
Elbow

Quadratus lumborum
Spinal extensors
Spine
Rectus abdominis
Serratus anterior
Pectoralis major

KEY
- •-- *Joints*
- ○— *Muscles*
- ● Engaging
- ● Engaging while stretching
- ● Stretching

Torso
Your **spinal extensors** engage while your **abdominals** stretch. Your **pectoralis major** stretches. Your **middle** and **lower trapezius** engage with your **rhomboids** to retract your scapulae, while your **serratus anterior muscles** engage to help stabilize them.

» CLOSER LOOK

Locust strengthens the entire back of your body, which can be particularly helpful for improving posture and core function. You do not have to lift yourself very far off the ground to gain the benefits of this pose.

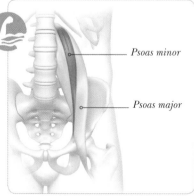

Psoas minor

Psoas major

Psoas minor

You'll probably feel your psoas stretching in this pose. Approximately 40 per cent of people have a psoas minor. This is further evidence of the variation between individuals; some people have more muscles or bones than others. Bodies are so different that, of course, each person's expression of a yoga pose will look unique.

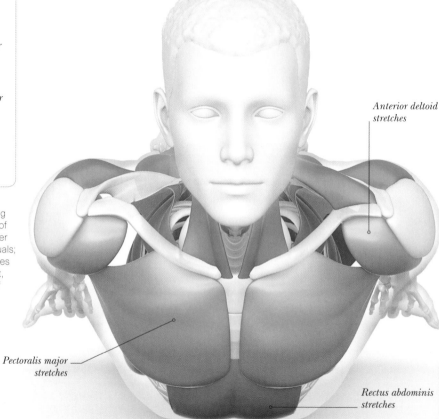

Crown of your head reaches up at a diagonal

Anterior deltoid stretches

Pectoralis major stretches

Rectus abdominis stretches

ANTERIOR VIEW

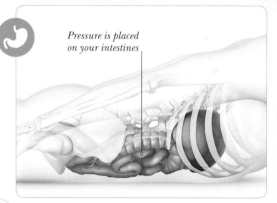

Pressure is placed on your intestines

Boost digestion

Poses like this can help stimulate a bowel movement because of the floor's pressure on your digestive organs, and via engagement of your core. This effect may be enhanced if you come in and out of the pose several times, simulating the rhythmic movements of your intestines.

Feet are pointed (in plantar flexion)

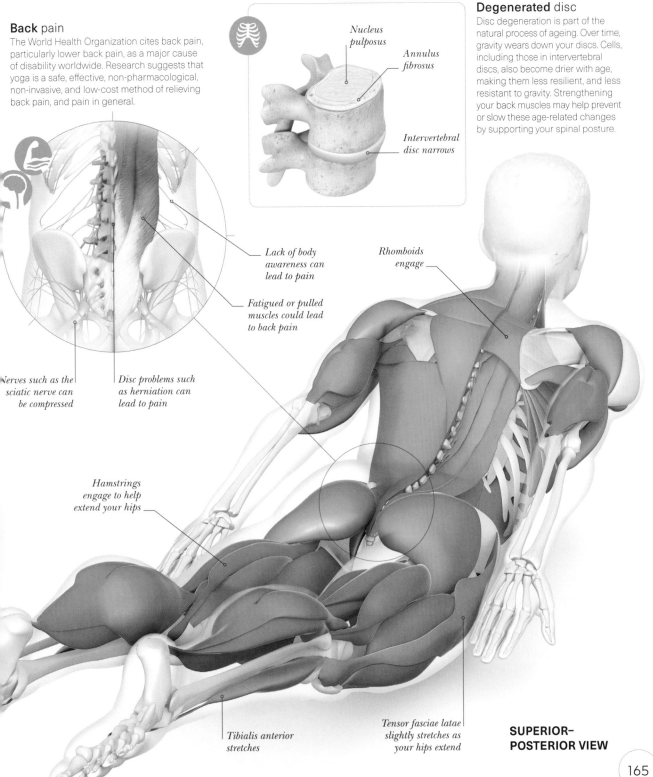

Back pain

The World Health Organization cites back pain, particularly lower back pain, as a major cause of disability worldwide. Research suggests that yoga is a safe, effective, non-pharmacological, non-invasive, and low-cost method of relieving back pain, and pain in general.

Nucleus pulposus

Annulus fibrosus

Intervertebral disc narrows

Degenerated disc

Disc degeneration is part of the natural process of ageing. Over time, gravity wears down your discs. Cells, including those in intervertebral discs, also become drier with age, making them less resilient, and less resistant to gravity. Strengthening your back muscles may help prevent or slow these age-related changes by supporting your spinal posture.

Rhomboids engage

Lack of body awareness can lead to pain

Fatigued or pulled muscles could lead to back pain

Nerves such as the sciatic nerve can be compressed

Disc problems such as herniation can lead to pain

Hamstrings engage to help extend your hips

Tibialis anterior stretches

Tensor fasciae latae slightly stretches as your hips extend

SUPERIOR-POSTERIOR VIEW

SUPINE LEG STRETCH

Supta Padangusthasana

This pose and its variations stretch your thighs in a way that is particularly safe for your lower back. This can be very relaxing and great for winding down after a long day. If you are unable to grasp your toes, try holding onto a strap around the bottom of your foot.

THE **BIG** PICTURE

The back of your lifted thigh and leg intensely stretch. Your arms gently pull your leg in, but you should try to relax any muscles that are not necessary for this action (like your jaw, neck, and shoulders).

ALIGNMENT
Your spine is neutral, or your lower back may be slightly flexed depending on how far into the pose you go. Pull your toe in until you feel a deep but comfortable stretch in your hamstrings.

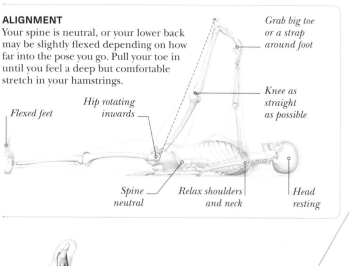

Flexed feet

Hip rotating inwards

Grab big toe or a strap around foot

Knee as straight as possible

Spine neutral

Relax shoulders and neck

Head resting

Lifted **thigh** and lower **leg**
Your **hip flexors** engage while your **quadriceps** extend your knee. Your **hip extensors** – particularly your **hamstrings** and **gluteus maximus** – stretch. As you grasp at your toes you'll likely feel your **ankle plantar flexors** – especially your **calf muscles** – stretching.

Lowered **thigh** and lower **leg**
In this version of the pose, your lowered thigh and leg are slightly engaged to stabilize. Your **hip flexors** are in a slightly lengthened position, your **quadriceps** extend your knees, and your **hamstrings** are slightly engaging. Your **ankle dorsiflexors** engage while your **plantar flexors** are in a neutral or lengthening position.

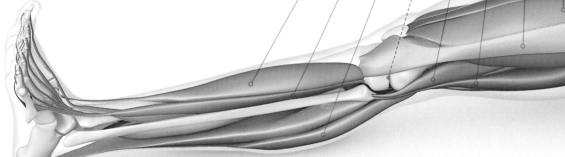

Tibialis anterior

Extensor digitorum longus

Gastrocnemius

Knee

Biceps femoris (short head)

Biceps femoris (long head)

Iliotibial band

Vastus lateralis

Rectus femoris

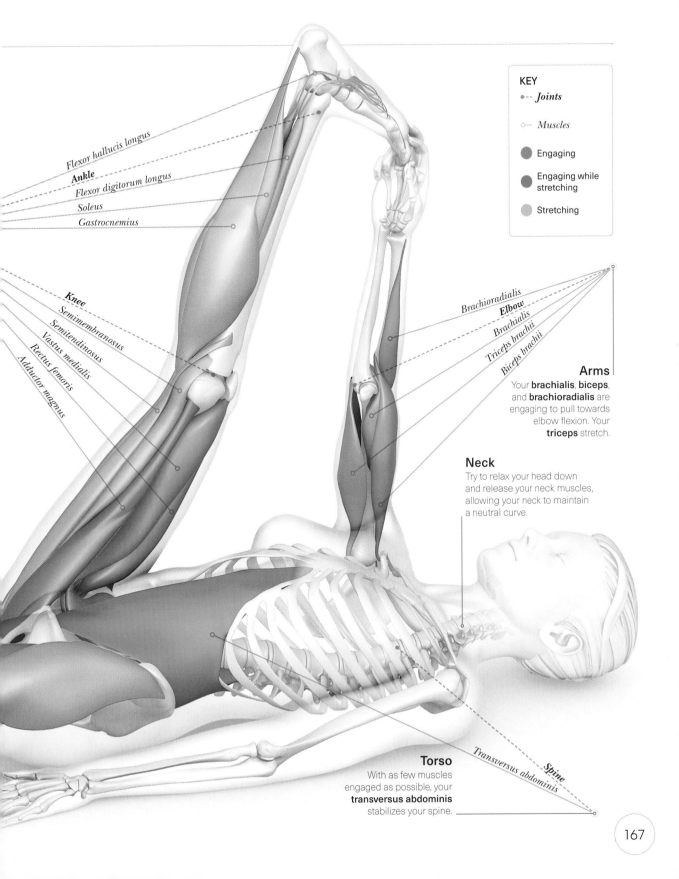

Flexor hallucis longus

Ankle

Flexor digitorum longus

Soleus

Gastrocnemius

Knee

Semimembranosus

Semitendinosus

Vastus medialis

Rectus femoris

Adductor magnus

KEY

•--- *Joints*

○— *Muscles*

● Engaging

● Engaging while stretching

● Stretching

Brachioradialis

Elbow

Brachialis

Triceps brachii

Biceps brachii

Arms

Your **brachialis**, **biceps**, and **brachioradialis** are engaging to pull towards elbow flexion. Your **triceps** stretch.

Neck

Try to relax your head down and release your neck muscles, allowing your neck to maintain a neutral curve.

Torso

With as few muscles engaged as possible, your **transversus abdominis** stabilizes your spine.

Transversus abdominis

Spine

167

» **CLOSER** LOOK

This stretch can be done with or without a strap, making it accessible for many people. Use your neurophysiology to your advantage to get a more effective stretch with mindfulness tricks.

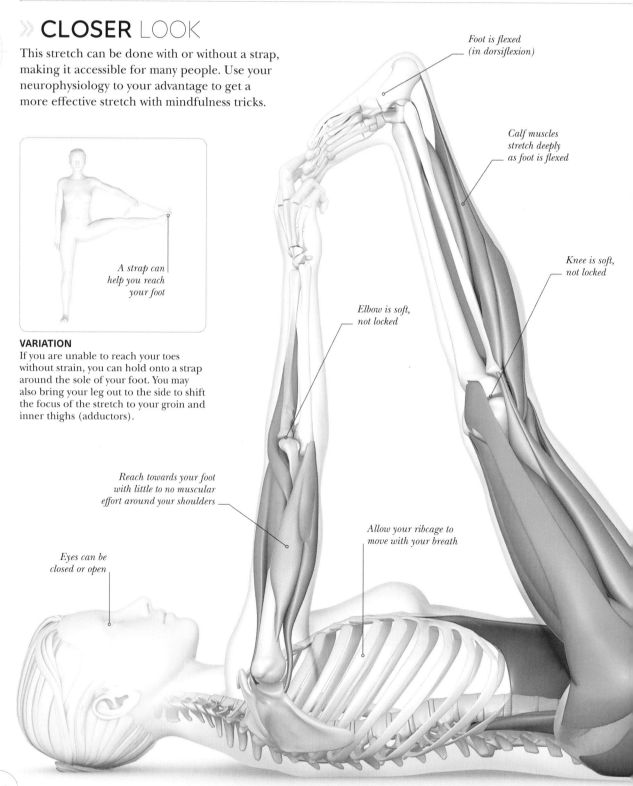

A strap can help you reach your foot

VARIATION
If you are unable to reach your toes without strain, you can hold onto a strap around the sole of your foot. You may also bring your leg out to the side to shift the focus of the stretch to your groin and inner thighs (adductors).

Foot is flexed (in dorsiflexion)

Calf muscles stretch deeply as foot is flexed

Knee is soft, not locked

Elbow is soft, not locked

Reach towards your foot with little to no muscular effort around your shoulders

Allow your ribcage to move with your breath

Eyes can be closed or open

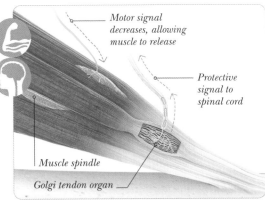

Release

When you first go into a stretching pose, you probably feel a taut pulling in your muscles. After a few breaths, tension peaks, and sensors in tendons called the Golgi tendon organ send a protective signal, inhibiting contraction and resistance in larger muscle fibres. This causes that pleasurable "ahhh" feeling of release.

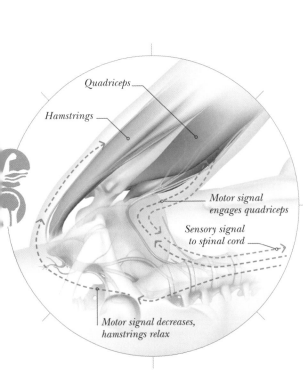

Reciprocal inhibition

Muscles often work in pairs. You can use reciprocal inhibition (RI), a protective physiological phenomenon, to get a deeper stretch safely. To initiate RI, consciously engage your quadriceps for a few breaths. Nerves in your quadriceps send a message to the paired hamstrings, telling them to relax further into the stretch.

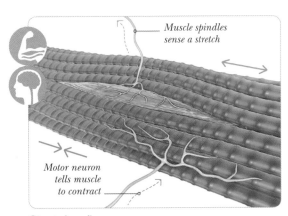

Stretch reflex

Smaller muscle fibres with sensors, called muscle spindles, don't release as soon, causing the stretch reflex (which involves muscle contraction to protectively resist overstretching). Override this by moving gradually into the pose, allowing muscle fibres to slowly release, to get a deeper stretch without injury.

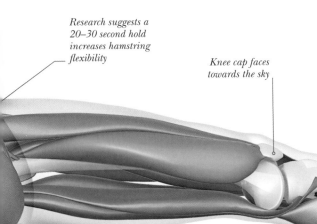

LATERAL VIEW

SUPINE TWIST
Supta Matsyendrasana

This relaxing spinal twist is often done at the end of a yoga class to calm your nervous system. Cultivate a sense of groundedness by releasing your body weight down into the floor. Find ease to activate the rejuvenating "rest and digest" part of your nervous system.

THE BIG PICTURE

This pose stretches muscles along your spine, including the small muscles that rotate it. Your shoulders, glutes, and thigh muscles are also stretching, though elsewhere in your body your muscles should be as relaxed as possible.

ALIGNMENT

Release completely to gravity, feeling your bones dropping down. If your shoulders or knee can't completely release, feel free to use a blanket or bolster for support.

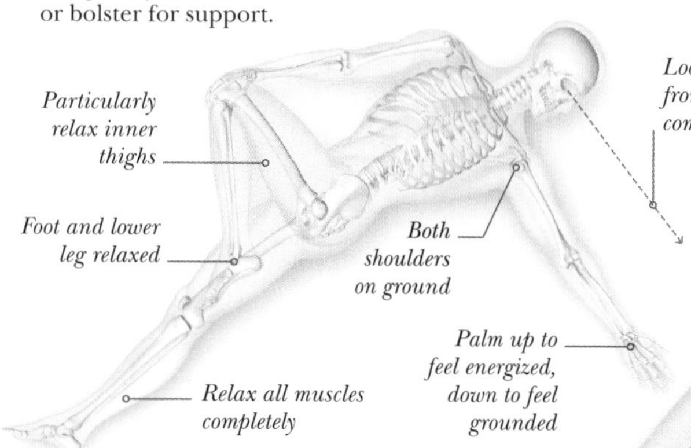

Particularly relax inner thighs

Foot and lower leg relaxed

Relax all muscles completely

Both shoulders on ground

Palm up to feel energized, down to feel grounded

Look away from knees if comfortable

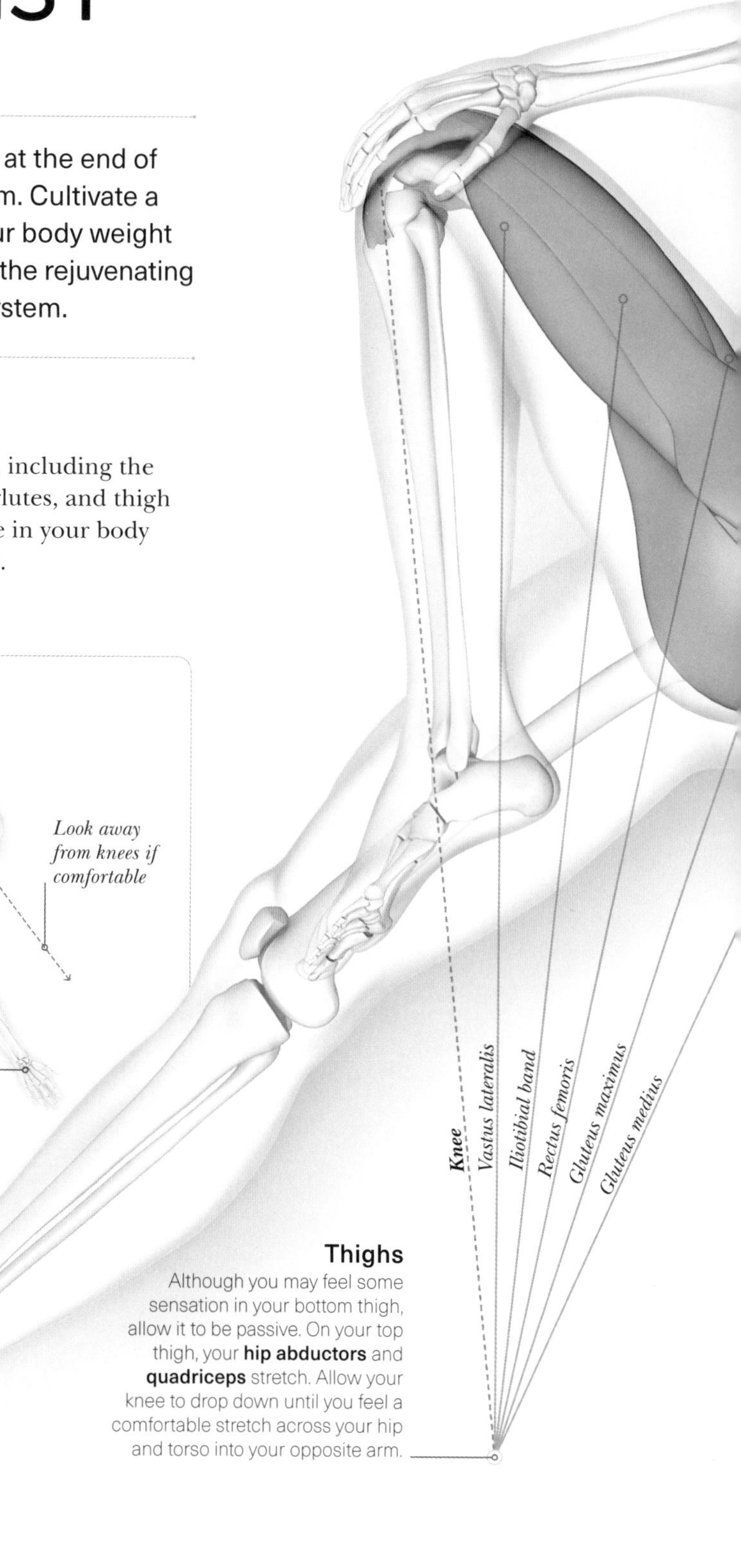

Knee

Vastus lateralis

Iliotibial band

Rectus femoris

Gluteus maximus

Gluteus medius

Thighs

Although you may feel some sensation in your bottom thigh, allow it to be passive. On your top thigh, your **hip abductors** and **quadriceps** stretch. Allow your knee to drop down until you feel a comfortable stretch across your hip and torso into your opposite arm.

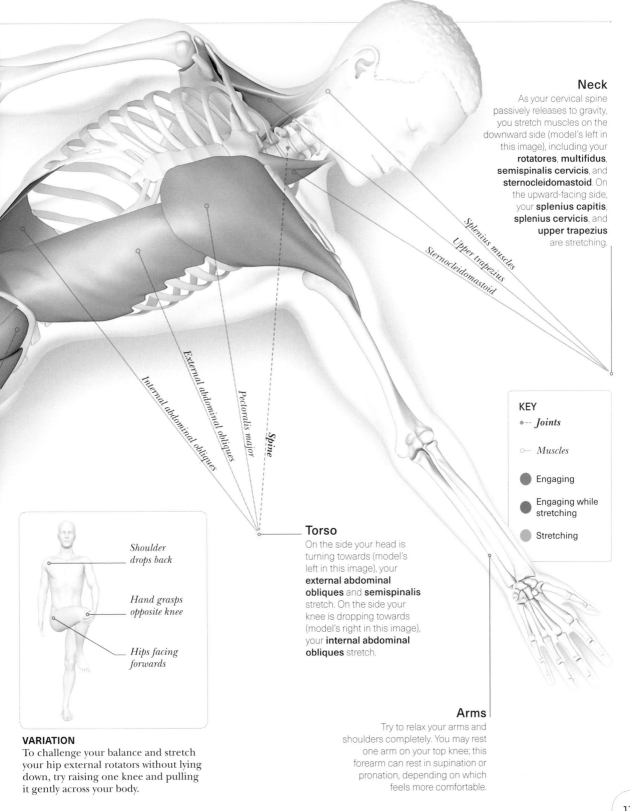

Neck

As your cervical spine passively releases to gravity, you stretch muscles on the downward side (model's left in this image), including your **rotatores**, **multifidus**, **semispinalis cervicis**, and **sternocleidomastoid**. On the upward-facing side, your **splenius capitis**, **splenius cervicis**, and **upper trapezius** are stretching.

Splenius muscles

Upper trapezius

Sternocleidomastoid

Internal abdominal obliques

External abdominal obliques

Pectoralis major

Spine

KEY

- •--- *Joints*
- ○— *Muscles*
- ● Engaging
- ● Engaging while stretching
- ● Stretching

Torso

On the side your head is turning towards (model's left in this image), your **external abdominal obliques** and **semispinalis** stretch. On the side your knee is dropping towards (model's right in this image), your **internal abdominal obliques** stretch.

Arms

Try to relax your arms and shoulders completely. You may rest one arm on your top knee; this forearm can rest in supination or pronation, depending on which feels more comfortable.

Shoulder drops back

Hand grasps opposite knee

Hips facing forwards

VARIATION

To challenge your balance and stretch your hip external rotators without lying down, try raising one knee and pulling it gently across your body.

»CLOSER LOOK

For many, Supine Twist is a safe way to do spinal rotation with ease. Wiggle into the pose and use props like a blanket until you find a pain-free position.

Spine safety

Supine Twist can be safer than seated or standing twists by changing the orientation of the impact of gravity on your intervertebral discs and spine. Also, spinal flexion often occurs with upright twists and the combination of rotation and flexion increases the risk of spinal issues.

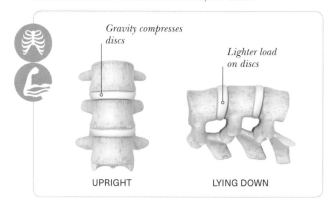

Gravity compresses discs

Lighter load on discs

UPRIGHT

LYING DOWN

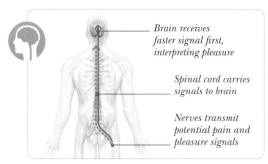

Brain receives faster signal first, interpreting pleasure

Spinal cord carries signals to brain

Nerves transmit potential pain and pleasure signals

Perceived pain pathway

Imagine two signals like trains simultaneously travelling to your brain: the red train pathway carries a signal that could be perceived as painful (nociceptive), and the green train pathway carries a signal that could be perceived as pleasurable. The green train is faster, reaching your brain first, possibly overriding nociceptive signals. This is called the gate theory of pain.

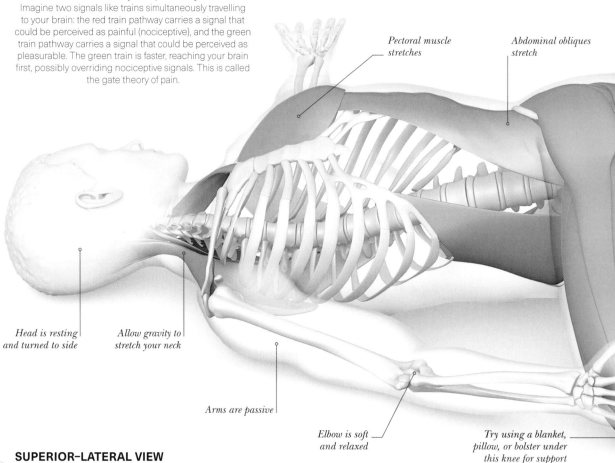

Pectoral muscle stretches

Abdominal obliques stretch

Head is resting and turned to side

Allow gravity to stretch your neck

Arms are passive

Elbow is soft and relaxed

Try using a blanket, pillow, or bolster under this knee for support

SUPERIOR-LATERAL VIEW

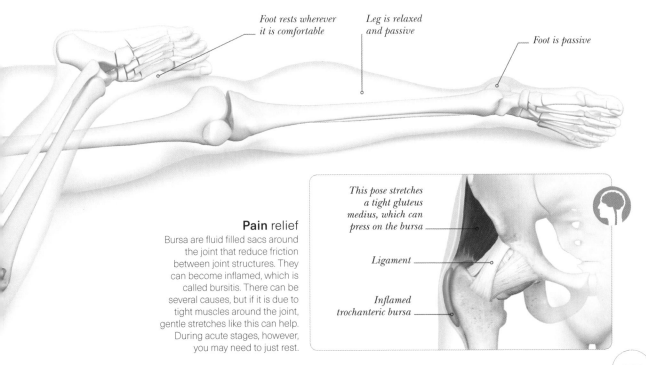

| | LATERAL MOVEMENT | FLEXION AND EXTENSION | ROTATIONAL MOVEMENT |

Cervical (Oc–T1)

Cervical spine allows most lateral movement

Thoracic (T1–L1)

Greatest range at base of spine

Thoracic spine has variable capacity for rotation

Lumbar (L1–S1)

Degrees (0°)   7.5°   0°   7.5°   10°   0°   10°   23.5°   10°   0°   10°   23.5°

LATERAL MOVEMENT   FLEXION AND EXTENSION   ROTATIONAL MOVEMENT

Spinal motion

Notice that your cervical and thoracic spine allow more twisting action than the lumbar. The shape of your vertebrae in each area facilitates or limits the amount of movement. Technically, you won't have a perfectly even twist. That is a visualization to help prevent extreme mobility or pinching in any one area. Different segments of your spine allow varying amounts of other motion.

Foot rests wherever it is comfortable

Leg is relaxed and passive

Foot is passive

Pain relief

Bursa are fluid filled sacs around the joint that reduce friction between joint structures. They can become inflamed, which is called bursitis. There can be several causes, but if it is due to tight muscles around the joint, gentle stretches like this can help. During acute stages, however, you may need to just rest.

This pose stretches a tight gluteus medius, which can press on the bursa

Ligament

Inflamed trochanteric bursa

173

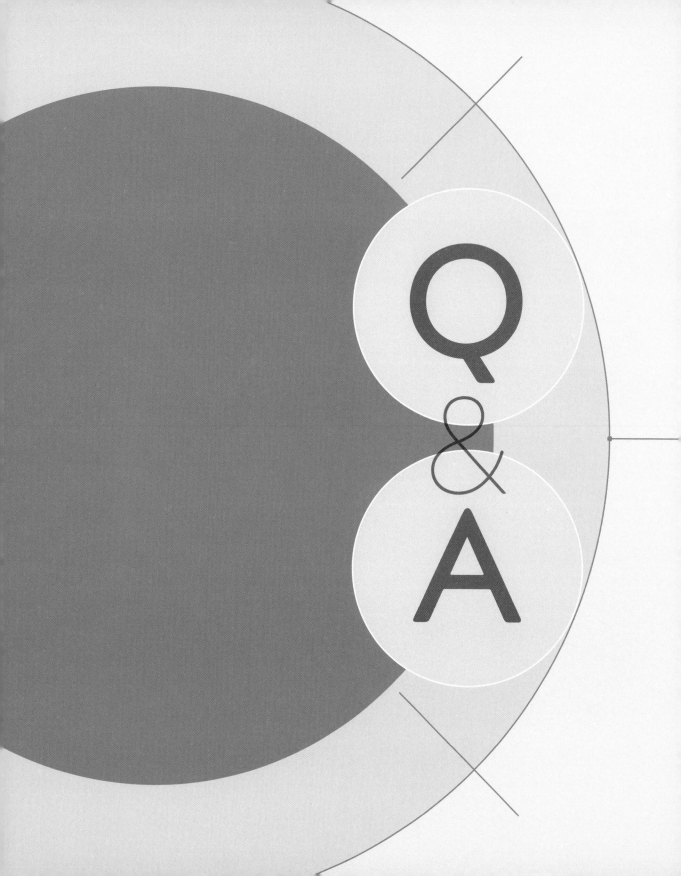

QUESTIONS AND ANSWERS

These Q&As are based on common questions I have had from my students over the years. The physical body is addressed first, then mental and more subtle layers of self. It is important to note that although yoga is based in Hindu traditions, its practices and wisdom are adaptable for everyone. Whether you are spiritual, religious, agnostic, or something else, yoga can help you find health and peace.

JOINTS AND FLEXIBILITY

While a certain degree of flexibility is important in accomplishing many asanas and completing daily activities, it's crucial to understand your body and know your limits so you can avoid injury and look after your joints. If you are very flexible, it may be best to focus on strengthening asanas.

❝❞

Yoga has been widely shown to increase flexibility so a lack of flexibility only gives you more reason to practise

Most of the
360
joints in the body are synovial, or free-moving

Q CAN I DO YOGA IF I'M NOT FLEXIBLE?

Yes. Yoga has been widely shown to increase flexibility, so a lack of flexibility only gives you more reason to practise. If you have limited range of motion (ROM) in a pose because your muscles are tight or you are recovering from an injury, it can be helpful to visualize your body moving further into the pose. Research suggests that this creates a neural map, sending signals to the muscles which leads to increased ROM. Similarly, research has found that visualizing yourself doing a pose and getting stronger can measurably strengthen your muscles, even without moving.

Q WHY DO MY JOINTS "POP"?

Most joints have synovial fluid between the bones, which contains dissolved gas molecules. Creating more space in the joint – for example, by pulling your thumb – pulls gases out of the fluid, similar to how CO_2 bubbles fizz out of carbonated drinks when you open the bottle. The gases re-dissolve into the fluid, and can be "popped" again after 20–30 minutes. There is no evidence to suggest this causes arthritis, but it may make your joints larger. If your joints pop with no wait, the joint structures may be rubbing against each other. This could slowly damage the joint structures.

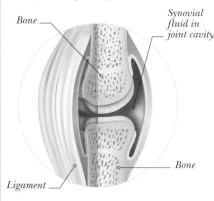

Bone

Synovial fluid in joint cavity

Bone

Ligament

SYNOVIAL JOINT

Q IS IT POSSIBLE TO STRETCH TOO MUCH?

Yes. There is a correlation between hypermobility – the ability to stretch beyond the normal range, or being "double jointed" – and chronic joint pain. When you stretch, you should feel the stretching sensation in the centre of the muscle, not near the joints, and you want to be able to breathe smoothly through the stretch. If you feel sharp or shooting sensations, numbness, pain, or anything that makes you grimace or hold your breath, you are overstretching. Overstretching lengthens your ligaments and/or tendons and, since they don't have much elasticity, they don't recoil well after they have been stretched. In other words, when the stress (load or stretch) on the tissue reaches the yield point it stops being "elastic" and becomes "plastic" (see above right). In clinical terms, this represents a tear. To

avoid injury, it's best to strike a balance between using your yoga asana practice to improve your strength and using it to improve your flexibility.

Stress–strain curve

This graph shows how much stress your tissue (muscle, tendon, or ligament) can take before injury. In the elastic region, the tissue can still return to its normal length when the stress is removed, but in the plastic region, it can't recoil. The ultimate fail point is a complete tear. To avoid injury, don't push beyond your limits.

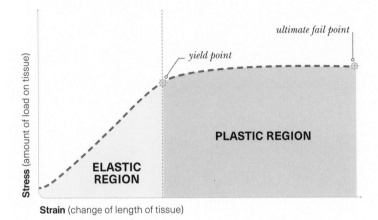

Stress (amount of load on tissue) / **Strain** (change of length of tissue)

ultimate fail point

yield point

PLASTIC REGION

ELASTIC REGION

MYTH-BUSTER

Hot yoga makes me more flexible.

It does, but only in the moment; it doesn't necessarily affect how flexible you are afterwards. Higher temperatures raise your metabolic rate, warming your tissue quicker so you can stretch deeper. Practising in hotter conditions makes it easy to stretch beyond your muscles' natural lengthening, which can lead to muscle damage (see above). Move slowly into poses with awareness to prevent injury.

SPINAL CARE

Your spine supports your whole body and protects your spinal cord, so looking after it is crucial for your health and wellbeing. Yoga helps care for your spine by encouraging good posture and alignment, but you may need to make simple adjustments to prevent or manage specific conditions and diseases.

Leaning forward over a smartphone can increase the load on the neck by **5** times

Q I SUFFER FROM NECK PAIN FROM TEXTING AND TYPING. CAN YOGA HELP?

Yes. While typing or texting, many of us allow our heads to fall forward. This increases the load on the neck and upper back muscles. With sustained strain, these muscles become inflamed and excessively tight, which can lead to pain. Yoga improves your awareness of how you hold your head throughout the day, which can prevent tech neck. To counteract its effects, you can also strengthen key muscles of proper neck posture by pressing your head back into your hands, a wall, or a car headrest for several breaths.

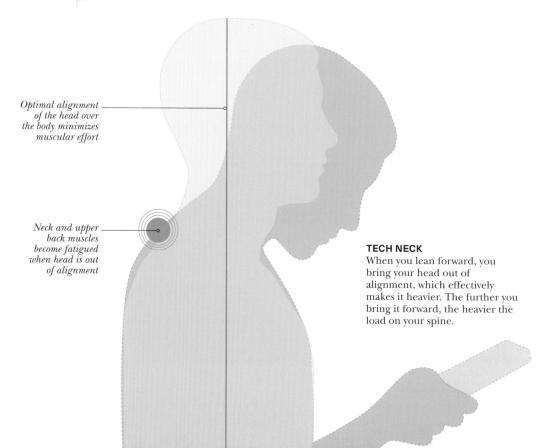

Optimal alignment of the head over the body minimizes muscular effort

Neck and upper back muscles become fatigued when head is out of alignment

TECH NECK
When you lean forward, you bring your head out of alignment, which effectively makes it heavier. The further you bring it forward, the heavier the load on your spine.

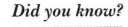

MYTH-BUSTER

I have back problems so I can't do yoga.

Research suggests that yoga is safe and effective for relieving chronic back pain. However, you may need to make adjustments to certain asanas or avoid some poses completely if you are managing a specific back condition (see pp.202–205). For many people, for example, touching the floor in Standing Forward Fold is not possible or comfortable, particularly for the lower back (the lumbar spine). However, you can still get the main benefits of the pose by bringing the floor closer to you, for example, by resting your hands on a block or on the base of a chair.

> *Research suggests that yoga is **safe** and **effective** for relieving chronic back pain*

Q IS THERE AN ALTERNATIVE TO ROLLING UP FROM A STANDING FORWARD FOLD?

The cue of rolling up from a Standing Forward Fold "vertebra by vertebra" is likely to have come from the dance world. Biomechanically and functionally, this transition has more risks than benefits. For many, it feels good and improves coordination. However, rolling up could lead to or exacerbate a herniated disc or a spinal fracture for those with osteoporosis. This transition also doesn't prepare you properly for real-world activities, such as picking things up safely. To avoid potential injury, and to build the muscle memory of safe movement patterns, try coming out of a Standing Forward Fold in the following way:

1 **Create a wider base** of support with your toes turned out slightly. This reduces the pressure on your knees.

2 **Bring your hands to** your hips or the front of your thighs.

3 **Keeping a neutral spine,** engage your core and push up to standing, as with a hip hinge. This can particularly recruit your transversus abdominis which may help alleviate lower back pain.

Did you know?

BACK PAIN IS ONE OF THE MOST COMMON **DISABLING AILMENTS** AND IS A LEADING CAUSE OF **LOST PRODUCTIVITY**. RESEARCH SHOWS YOGA NOT ONLY **REDUCES BACK PAIN** BY CLINICALLY SIGNIFICANT LEVELS, BUT ALSO **REDUCES** THE NUMBER OF SICK DAYS TAKEN.

LIFE STAGES

Not only is it possible and safe to practise yoga during different life stages – from childhood to pregnancy to old age – but research is now building to show that yoga and its accompanying practices, such as meditation, can bring additional benefits at these times of life.

"

*Yoga emphasizes the **whole child**, so it fulfils an important need for **social and emotional** learning*

There are over
900
"yoga in school" programmes in North America

Did you know?

RESEARCH SUGGESTS THAT YOGA COULD **IMPROVE** CORE SYMPTOMS OF ATTENTION DEFICIT HYPERACTIVITY DISORDER (ADHD) SUCH AS INATTENTION, HYPERACTIVITY, AND IMPULSIVITY, IN **CHILDREN** AND ADOLESCENTS, WHEN PRACTISED AS A **MIND-BODY** THERAPY AND FORM OF EXERCISE.

Q DOES YOGA BENEFIT CHILDREN?

A focus on academic performance can result in children sitting for long periods of time and can lead to other vital life skills being overlooked. As a holistic practice, yoga emphasizes the whole child, so it fulfils an important need for social and emotional learning (SEL). Yoga can affect all components of the social and emotional learning model, which include:

● **Self-awareness:** recognizing and identifying emotions
● **Self-management:** regulating emotions and managing stress
● **Social awareness:** acknowledging the perspectives of others
● **Relationship skills:** creating and maintaining a social network
● **Responsible decisions:** making conscious, positive decisions.

A review of research from Harvard and the Kripalu Center for Yoga and Health, for example, found that using yoga therapeutically was a viable way to improve the physical and mental health of children and adolescents. Meditation programmes in schools have also shown strong improvements in resilience to stress and cognitive performance.

Q IS YOGA SAFE DURING PREGNANCY? DOES IT HAVE ANY BENEFITS?

Yes. Prenatal yoga classes are widely available and are often recommended by doctors. Research, including a 2015 study from Alpert Medical School of Brown University, has suggested that prenatal yoga is not only safe for both the expectant mother and the baby (as measured by fetal heart rate), but it can also be beneficial for the fetus and mother while pregnant, throughout labour and delivery, and postpartum. Small studies have also suggested that, during pregnancy, yoga may have the positive effects shown below.

REDUCES
- pelvic pain and overall pregnancy discomfort
- stress, depression, and anxiety signs
- postpartum depression.

IMPROVES
- optimism, empowerment, wellbeing, and social support
- birth weight (by reducing risk of pre-term labour).

> *Meditation may **slow** or even **prevent** some of the natural degradation of **brain tissue** that happens with aging*

Q HOW DOES MEDITATION AFFECT MY BRAIN AS I AGE?

Many areas of your brain tend to shrink with age, but Harvard neuroscientist Sara Lazar, PhD, and her team have shown via MRI brain scans that 50-year-old meditators have key brain structures similar to that of 25-year-old non-meditators. This suggests that meditation may slow or even prevent some of the natural degradation of brain tissue that happens with aging. This is thanks to neuroplasticity (see pp.26–27). While it is likely that other factors are involved, such as lifestyle and diet, it is feasible that meditation and the resulting mindset contribute significantly. Research also suggests that your brain can start to make these changes in eight weeks. A daily 30-minute mindfulness practice (including a body scan, yoga, and seated meditation practice) has been shown to change the brain in ways that result in better memory and improved problem-solving. A mindfulness questionnaire also showed that eight weeks' instruction and practice improved three key qualities that may contribute to a positive mindset as we age: observing internal and external environment; acting with awareness instead of reacting; and the non-judgment of inner experience.

Continued →

66 99

*The yogic concept of **equanimity** teaches us to handle **change** and challenges with **grace***

Q HOW DOES YOGA AFFECT HOW WE AGE?

According to experts, yoga has the following benefits that support healthy aging:

- builds muscle strength to counteract the natural skeletal atrophy that happens with aging
- improves flexibility to prevent the loss of range of motion
- improves dynamic and static balance, reducing your risk of falling
- improves mental and physical agility so you can react faster.

Yoga improves strength, flexibility, balance, and agility in both physical and mental realms. Together, all of this may help improve your healthspan – the number of years you live without illness.

Did you know?

IN 2050, **ONE FIFTH** OF THE WORLD'S POPULATION WILL BE **AGED 60** OR OVER. THIS MAKES IT **MORE IMPORTANT** THAN EVER TO PREPARE OUR BODIES FOR **HEALTHY AGING** WITH PRACTICES SUCH AS YOGA.

MYTH-BUSTER

I'm too old to practise yoga.

Studies of yoga and the elderly have shown improvements in flexibility, strength, balance, and functional activities, such as getting up from a chair. Yoga is also highly customizable. You can practise simple breathwork and adapt any asana, for example using a chair, blocks, or blankets.

Q CAN YOGA HELP ME KEEP MY INDEPENDENCE?

Yes. Practising yoga can help you maintain independence by preserving functional abilities so you can perform daily activities and continue doing what you love. Applying the philosophy of yoga to your life can also help you find purpose and meaning, which contributes to independence and wellbeing. For example, the yogic concept of equanimity (mental calmness) teaches us to handle change and challenges with grace.

Just **8** weeks of mindfulness practice could slow brain changes associated with aging

Q HOW DOES YOGA AFFECT MY BONES AS I AGE?

Yoga can feasibly protect you from fractures associated with osteoporosis by preventing falls and strengthening the bone and muscles around common fracture sites, such as T9 and T10 (vertebrae at the base of the upper back), wrists, and the hip, particularly with asanas such as the below. Yoga also helps maintain the ability to safely get up and down from the floor so you can protect your joints and keep active.

Yoga is becoming more popular with those aged **65** and over

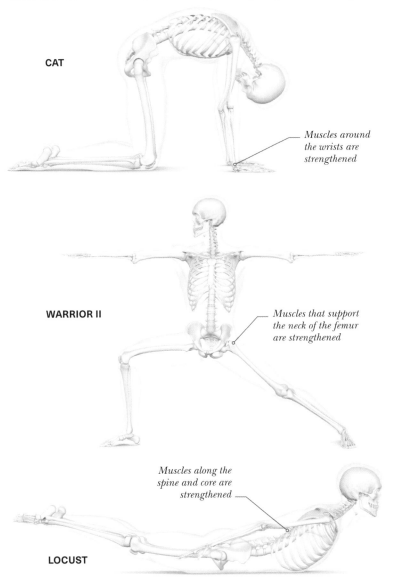

CAT

Muscles around the wrists are strengthened

WARRIOR II

Muscles that support the neck of the femur are strengthened

Muscles along the spine and core are strengthened

LOCUST

MEDITATION

Yoga was traditionally seen as a way to prepare the body for meditation. Today, many yoga classes include meditative elements, such as mindfulness practices and chanting "om", as ways to relax the body and mind. Science shows that the benefits of these meditative practices also extend into your daily life.

" "

*Simply **observe** your thoughts arising. It's like **watching clouds** pass by while remaining aware of the vast, clear blue sky*

Q IS MINDFULNESS THE SAME AS MEDITATION? HOW IS IT PRACTISED?

Mindfulness is a simple and popular type of meditation that is often practised in traditional seated poses. It also refers to a mindset that you can bring into the rest of your life. According to Jon Kabat-Zinn, PhD, founder of the well-researched Mindfulness-Based Stress Reduction (MBSR) programme, mindfulness can be defined as deliberately paying attention to the present moment without judgment. It often involves observing breath, thoughts, sounds, or physical sensations, all of which are encouraged in yoga practice.

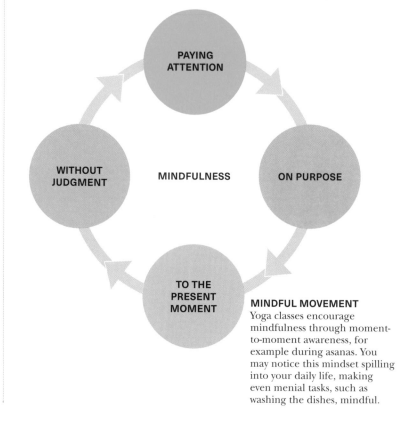

PAYING ATTENTION

MINDFULNESS

WITHOUT JUDGMENT

ON PURPOSE

TO THE PRESENT MOMENT

MINDFUL MOVEMENT
Yoga classes encourage mindfulness through moment-to-moment awareness, for example during asanas. You may notice this mindset spilling into your daily life, making even menial tasks, such as washing the dishes, mindful.

Q DOES MINDFULNESS REALLY WORK?

Anatomical MRI scans have shown changes in subjects' brain grey matter concentration after they participated in an eight-week MBSR programme, suggesting that MBSR affects areas of the brain involved in learning and memory processes, emotion regulation, self-awareness, and new perspective taking. Another study showed that even brief training in mindfulness reduced fatigue and anxiety, while longer training seems to particularly improve attention and focus.

Q HOW DO I SIT COMFORTABLY FOR MEDITATION?

Sitting on a cushion, folded blanket, pillow, or bolster helps you to elevate your hips at an angle and tilt your pelvis to neutral, bringing a natural inward (lordotic) curve to your lumbar spine. Another traditional meditation posture is Hero Pose (Virasana), or kneeling. If you feel any pain in your knees, you can use blocks or a bolster to elevate your hips. If neither of these positions work for you, you can also sit in a chair to meditate. Try to sit tall and forward in the chair, without leaning back. It may also help to sit on a cushion as this will tilt your pelvis forward slightly. Place your feet directly under your knees or a little ahead of them. If meditating in any of these seated positions is too uncomfortable, meditate in Savasana (see p.186).

(see p.186)

Did you know?

EXTREME FOCUS, SUCH AS WHILE PLAYING AN INSTRUMENT, HAS BEEN CONNECTED WITH MEDITATION. PSYCHOLOGISTS CALL THIS A "FLOW STATE". IN BOTH MEDITATION AND "FLOW", YOUR BRAIN WAVES CHANGE FROM BETA – ASSOCIATED WITH THINKING AND CONVERSING – TO MOSTLY ALPHA AND THETA – ASSOCIATED WITH RELAXATION AND CREATIVE PROBLEM SOLVING.

Q MY MIND IS SO BUSY. DOES THIS MEAN I'M NO GOOD AT MEDITATING?

No. Many people think meditating is about "stopping" thoughts, but it isn't. In the form of meditation most commonly practised in the modern day, you simply observe your thoughts arising. It's like watching clouds pass by while remaining aware of the vast, clear blue sky in which they float. When meditating, your only task is to gently remind yourself to come back to the present in a state of observing.

Q WHY DO WE CHANT "OM"?

An elongated exhale turns on the relaxation response. One small study also found that chanting "om" deactivates parts of the emotional brain related to fear, compared to chanting "sssss", as seen in fMRI brain imaging. This suggests that "om" may have benefits beyond the elongated exhale.

SAVASANA

Also known as Corpse Pose, Savasana is the final relaxation pose often practised for 5–15 minutes at the end of yoga classes. It is also used for meditative practices, such as yoga nidra. While there is still more research to be done, Savasana has been used clinically for its relaxation benefits.

*Savasana activates the **parasympathetic nervous system,** and all the **profound benefits** of this relaxation response*

WHAT IS SAVASANA AND WHAT ARE ITS BENEFITS?

Savasana is practised lying flat on your back with your legs and arms relaxed symmetrically, palms face up. It can also be used for relaxation and as a meditation posture if sitting is uncomfortable or you are not well. Its many benefits include:

• activating the parasympathetic nervous system (PNS), and all the profound benefits of this relaxation response, including lowering blood pressure and slowing heart rate
• teaching muscles to relax effectively
• increasing heart rate variability, representing resilience.

6

weekly sessions of yoga nidra improved stress, muscle tension, and self-care

WHAT IS PROGRESSIVE MUSCLE RELAXATION?

Progressive muscle relaxation (PMR) involves squeezing and then releasing your muscles, often sequentially from head to toe, while in Savasana. This encourages neuromuscular connection, giving the body-mind clear examples of tension and release, which helps the body relax physically. Immediately after your muscle fibres contract, they have the capacity to lengthen or relax even more.

WHY IS THERE OFTEN A LONGER, GUIDED SAVASANA AT THE END OF CLASS?

This is a mindfulness practice called yoga nidra. Nidra means sleep, so think of it as a "yoga nap". A general intention of the practice is to remain alert to allow observation of the physiological effects of each stage of sleep. It is usually practised in Savasana for 15–30 minutes and has shown promising results in small studies for improving sleep, decreasing depression, and managing chronic pain.

Q DOES YOGA NIDRA PROVIDE THE SAME BENEFITS AS SLEEP?

Although it does seem to offer many of the same rejuvenating benefits, yoga nidra does not replace sleep. However, it does produce brain wave patterns similar to those of sleep (see below).

Brain frequency chart

BRAINWAVE	SLEEP STAGE	YOGA NIDRA STAGE	LEVEL OF CONSCIOUSNESS	CHARACTERISTICS
GAMMA	Fully awake	Not nidra	Conscious	High alertness (not well understood)
BETA	Fully awake	Initially when transitioning into the practice	Conscious	Thinking and talking
ALPHA	First stage of sleep	During body scan and relaxation	Conscious – gateway to the subconscious	Relaxation
THETA	Next stage of sleep	May be reached, likely later in the practice	Subconscious	Creative problem solving
DELTA	Deep dreamless sleep	May be reached but there is little to no evidence of this	Unconscious	Rejuvenation and intuition

Q LYING FLAT ON MY BACK IS UNCOMFORTABLE. WHAT CAN I DO?

Many people find Savasana uncomfortable, particularly for their backs. Try using a support under your knees or lying in a constructive rest position – raising your knees and placing the soles of your feet on the floor – to relieve tension in your lower back. This can also help stop you from falling asleep.

MYTH-BUSTER

Savasana prevents lactic acid build up.

No. Lactic acid, a waste product from muscle engagement, is broken down and removed by your liver within minutes after exertion. To reduce soreness, build the intensity of your asana practice over time. You can also rest the sore muscles by doing a more restorative class or working different muscle groups.

STRESS

Common sense tells us that yoga helps us to manage stress by promoting relaxation and holistic wellbeing. But understanding the science behind the calming power of yoga can empower us to take a more proactive approach to a less stressed life, which enables us to achieve more positive health outcomes.

Q HOW DOES STRESS IMPACT MY HEALTH? AND HOW DOES YOGA HELP?

We tend to think of all stress as bad, but healthy levels of positive stress – eustress – can help us perform at our best. However, too much negative stress is associated with mental health imbalances and chronic pain, along with many of the industrial world's major killers, including heart disease, stroke, and cancer. It's important to recognize that stress doesn't necessarily cause these diseases. Research suggests that the greatest predictor of whether you will suffer from these diseases or not is not how much stress you experience, but how you deal with and think about stress. Those who have more negative emotions amidst stress are more likely to experience negative health outcomes. Yoga is an effective tool for managing stress because it helps us regulate our emotional response to stressors by teaching us to become the observer of our thoughts and feelings, and through improving our mind-body connection (see right). As a result, yoga can lead to more positive health outcomes.

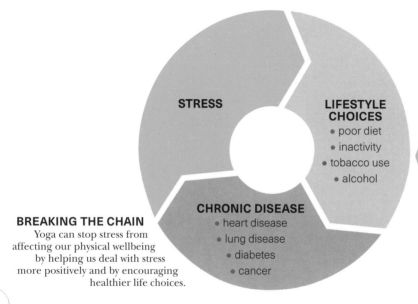

STRESS

LIFESTYLE CHOICES
- poor diet
- inactivity
- tobacco use
- alcohol

CHRONIC DISEASE
- heart disease
- lung disease
- diabetes
- cancer

Yoga helps us manage stress both in terms of how we view it and by activating the relaxation response and decreasing cortisol. Yoga practitioners are also more likely to make healthy lifestyle choices, such as exercising.

BREAKING THE CHAIN
Yoga can stop stress from affecting our physical wellbeing by helping us deal with stress more positively and by encouraging healthier life choices.

Q HOW DOES AN IMPROVED MIND-BODY CONNECTION HELP ME MANAGE STRESS?

Because yoga includes practices that engage both your mind and body, it helps you to regulate your system through both top-down and bottom-up pathways. Enhancing your mind-body and body-mind connections increases your ability to self-regulate and improves your resilience (your ability to bounce back after stress via homeostasis, the body's self-regulation of internal conditions). This all occurs partly due to the complex workings of your vagus nerve (see pp.190–91).

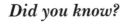

*Enhancing your mind-body and body-mind **connection** increases your ability to **self-regulate** and improves your **resilience***

NEUROCOGNITIVE (MIND-BODY) PATHWAY

1 Meditation, mindful movement, and intentional living based on the philosophical teachings of yoga increase your attention

2 Increased attention regulates your nervous system and helps you maintain homeostasis more efficiently

NEUROPHYSIOLOGICAL (BODY-MIND) PATHWAY

1 Yoga practices such as asanas, mudras, and pranayama, give you internal body awareness (interoception)

2 This interoceptive information affects your autonomic nervous system (ANS), which changes your thoughts and neural pathways, building your brain and improving self-regulation

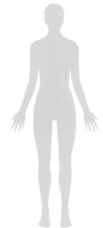

Did you know?

HANS SELYE COINED THE TERM **"STRESS"** IN 1936 TO DESCRIBE THE BODY'S **RESPONSE TO CHANGE**. HE IDENTIFIED TWO TYPES OF STRESS: **EUSTRESS**, WHICH IS BENEFICIAL STRESS, SUCH AS AN ENGAGING WORK PROJECT; AND **DISTRESS**, WHICH IS REAL OR IMAGINED STRESS THAT PUTS MORE PRESSURE ON YOUR SYSTEM.

Continued →

Q HOW DOES STRESS FIT INTO TRADITIONAL YOGIC PHILOSOPHY?

A 2018 article in *Frontiers in Human Neuroscience* aligns the ancient wisdom of yoga, particularly the gunas, with the role of the vagus nerve in our physiological response to stress and relaxation.

The vagus nerve is the only cranial nerve that leaves the head and neck area. It is mainly responsible for your relaxation response: telling your heart to slow, improving your digestion, and encouraging social connection. Rather than an "on/off" switch, the stress and relaxation responses work more like a dial, or dimmer knob. This allows adjustment to the perfect blend of electrical activity

from each branch of your autonomic nervous system (ANS) for the situation (see below).

According to the Polyvagal Theory proposed by American neuroscientist Stephen Porges, PhD, the vagus nerve is split functionally in a way that helps us adjust effectively. Researchers have explained this neural adaptability in terms of the gunas. Gunas means "thread" or quality. The three gunas – sattvic, rajasic, and tamasic – are the three essential aspects of nature that weave together to create what we observe as the reality of the material world (also known as prakriti) with its ever-changing conditions. Each of the gunas is associated with a state of mind and certain characterstics that map against the different functions of the vagus nerve (see below).

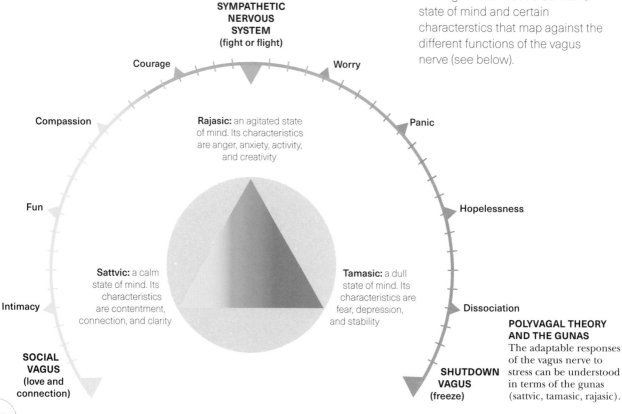

SYMPATHETIC NERVOUS SYSTEM (fight or flight)

Courage

Worry

Compassion

Panic

Rajasic: an agitated state of mind. Its characteristics are anger, anxiety, activity, and creativity

Fun

Hopelessness

Intimacy

Dissociation

Sattvic: a calm state of mind. Its characteristics are contentment, connection, and clarity

Tamasic: a dull state of mind. Its characteristics are fear, depression, and stability

SOCIAL VAGUS (love and connection)

SHUTDOWN VAGUS (freeze)

POLYVAGAL THEORY AND THE GUNAS
The adaptable responses of the vagus nerve to stress can be understood in terms of the gunas (sattvic, tamasic, rajasic).

SHOULD I BE CALM AND UNDER THE SOCIAL VAGUS OR SATTVIC STATE ALL THE TIME?

No. Yoga does teach our bodies to go into the sattvic state more often and more efficiently. This helps us to find balance in a world dominated by extremes of rajas and tamas. However, there is a misconception that yoga should make you perfectly calm all the time and that if that doesn't happen, you are bad at yoga. Constant calm is not the goal.

Your nervous system dynamically fluctuates, as do the gunas, throughout the day and over the course of your life to help you rise to the challenges your environment presents. Through yoga, you cultivate the capacity to be a non-judgmental observer of the constant changes so they don't control you. The ultimate ideal of this higher state of pure consciousness (also known as Purusha) is self-realization: finding meaning and connection amidst the experience of inevitable stressors. Increased consciousness of any level represents increased resilience.

*Through **yoga,** you cultivate the capacity to be a **non-judgmental observer** of the constant changes so they don't control you*

HOW CAN I RECOGNIZE AND REBALANCE THE NEGATIVE GUNAS?

The first step is to notice the signals of stress and the negative gunas in your body. These signals are different for everybody. Does your chest tighten or gut churn in an agitated, rajasic state? Do you tend to slouch or dissociate from sensations in a dull, tamasic state? Once you can recognize, identify, and observe your signals effectively, you can use the tools of yoga – including physical poses, mudras, breathwork, and meditation – to activate the relaxation response. Many yoga practices can be done discreetly throughout the day: no one will know you are elongating your exhales to calm down, adjusting your posture, or taking fuller breaths for more energy.

Did you know?

80 PER CENT OF THE VAGUS NERVE'S FIBRES SEND **INFORMATION** FROM THE BODY TO THE BRAIN. THIS MAKES IT A KEY PATHWAY OF **INTEROCEPTION** (INTERNAL BODY AWARENESS) FROM YOUR **HEART AND GUT** TO YOUR BRAIN. YOGA CAN IMPROVE YOUR INTEROCEPTION AND **VAGAL FUNCTION**.

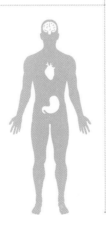

THE BRAIN AND
MENTAL WELLBEING

New research shows us that yoga changes how our brains work, for the better. Due to the neuroplasticity of our brains (see pp.26–27), these changes demonstrate the potential for yoga to become an effective adjunct to our medical and psychological care.

> *Yoga gives us the **tools** to break thought and **emotional patterns** that no longer serve us*

8 weeks of mindfulness meditation can help reduce fear-related activity in the brain

Q WHAT DOES YOGA DO TO MY BRAIN?

When your brain becomes accustomed to a well-worn neural path, it becomes a habit, such as mindlessly looking at your phone when you're bored. New neural paths can form in the same way, and repeated activation makes these paths bigger and stronger.

By reinforcing positive behaviours, yoga gives us the tools to break thought and emotional patterns that no longer serve us. This allows the choice of healthier patterns when challenges arise, making yoga a powerful practice for our mental health and wellbeing.

Q HOW CAN YOGA HELP MY MENTAL WELLBEING?

Sometimes we get stuck in a rajasic (the energy of agitation), reactionary pattern or a tamasic (the energy of resistance) slump. Yoga alone is not enough to manage a serious mental health concern, but it can be an effective supplement to your medical and psychological care because it affects how your brain responds to mental challenges.

In simplified terms, there are three structures within the brain:

- **The instinctual brain** (brain stem), which asks "Am I safe?"
- **The emotional brain** (limbic system), which asks "What am I feeling?"
- **The thinking brain** (frontal cortex), which asks "What does this mean?"

Under trauma, depression, chronic stress, or anxiety, you may have an overactive emotional brain. Signals from your amygdala (the "fear centre" of your emotional brain) encourage fight-or-flight responses from your instinctual brain, causing the stress response to override the relaxation response. When this happens often, your thinking brain is less effective at regulating. Yoga – including asanas, pranayama, and meditation – teaches the thinking brain to better regulate mood and emotional states amidst stressors in life (see p.188).

Q WHAT EVIDENCE IS THERE TO SHOW THAT YOGA REALLY CHANGES OUR BRAINS?

A number of studies have focused on this. One 2015 review of two decades of research found that specific areas of the brain are commonly affected by the yoga-based practice of mindfulness, as shown in the diagram, right. It showed that key areas of the frontal cortex are strengthened, helping you effectively recognize and regulate emotions. Brain scans reported in a different research article, from 2018, also demonstrated that yoga asanas and meditation both reduced amygdala volume on the right-hand side of the brain, which is more associated with negative emotions and fear. In addition, researchers at Stanford University found that eight weeks of mindfulness meditation enabled people to better reduce fear-related amygdala activity. This seems to work largely as a result of participants being mindful of sensations and emotions instead of pushing them down.

KEY

- Parts of the brain associated with **emotional regulation** are strengthened
- The striatum, associated with **emotional regulation and attention control**, is strengthened
- Anterior cingulate cortex, associated with **attention control**, is strengthened
- Parts of the brain associated with **self-awareness** are strengthened
- The amygdala, associated with **fear**, is reduced

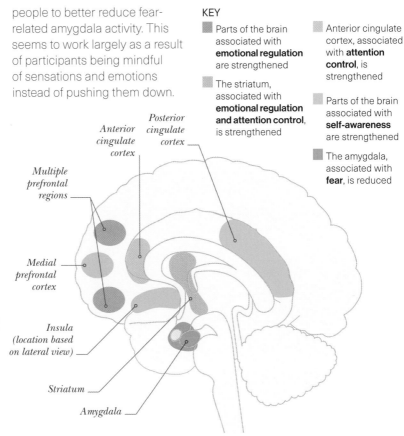

Posterior cingulate cortex

Anterior cingulate cortex

Multiple prefrontal regions

Medial prefrontal cortex

Insula (location based on lateral view)

Striatum

Amygdala

MIDSAGITTAL VIEW OF THE BRAIN

Did you know?

RESEARCHERS BELIEVE THAT **SOMATIC PRACTICES** (OR MOVEMENT PRACTICES THAT EMPHASIZE **PERCEPTION**, SUCH AS YOGA ASANAS) ARE USEFUL FOR HELPING PEOPLE TO **PROCESS TRAUMA** WITHOUT RE-TRIGGERING BECAUSE THEY HELP US **RELEASE TENSION** HELD IN THE BODY.

CHRONIC PAIN

Acute pain, such as an ankle sprain or a slip-and-fall injury, often needs rest to heal, which may mean avoiding or modifying yoga poses. But when pain becomes chronic, mind-body practices such as yoga have been shown to be well-suited to providing a safe supplement to medical care.

Q CAN YOGA REALLY HELP WHEN PAIN BECOMES CHRONIC?

Yes, there is evidence to show that it can help. Pain becomes chronic when it persists beyond the healing time of about three months. If you suffer from chronic pain, such as many cases of back pain or arthritis, you generally don't need to rest more because there may be little to no physical damage to heal. In fact, you probably need to move more because exercise tends to help relieve chronic pain, along with reducing associated stress.

Yoga practices have been shown to have an analgesic – or pain-relieving – effect. In one study of military veterans with lower back pain, opioid use declined in all subjects after a 12-week, twice a week yoga programme.

MYTH-BUSTER

Meditation relieves pain because of the placebo effect.

Recent research has shown that mindfulness meditation works better than a placebo in reducing pain. Subjects were exposed to a painful heat stimulus before and after receiving treatment: a placebo cream, "fake meditation", and traditional mindfulness meditation. The intensity and unpleasantness of the pain was evaluated psychophysically and by functional neuroimagery. The mindfulness group's pain intensity and unpleasantness reduced most significantly.

Four **20**-minute mindfulness classes can reduce pain's unpleasantness by **57%**

BRAIN PERCEIVES PAIN MORE FREQUENTLY

LESS LIKELY TO MOVE, SO PAIN IS NOT RELIEVED

BRAIN IS UNABLE TO INTERPRET/ REGULATE PAIN SIGNAL

EMOTIONAL AND PHYSICAL HEALTH IS AFFECTED

CHRONIC PAIN CYCLE
When the brain frequently perceives signals as pain, it becomes inured, and is unable to regulate its response. Yoga helps break the cycle.

Q WILL ASANA PRACTICE REDUCE MY CHRONIC PAIN?

It depends. Some asanas can help reduce pain by stretching and strengthening the affected area(s). However, biomechanics is just one piece of the puzzle. At its most basic level, what your brain interprets as "pain" starts as a signal received from a receptor (nociceptor) in your body. Research has shown that the amount of pain perceived doesn't depend on the amount of tissue damage as seen in X-ray or MRI scans. This means that without the brain, there is no pain; but this doesn't mean pain is imagined. Your brain builds your pain experience just as it constructs your reality and perspective. The level of pain you experience is based on your brain's interpretation of the level of danger those signals represent. So, as with chronic stress, chronic pain is partly a problem with regulation, often related to a faulty alarm system. Research shows that relaxing yoga asanas and practices, such as meditation and pranayama, can help regulate the pain response.

> *Relaxing **yoga asanas** and practices, such as **meditation** and pranayama, can help **regulate** the pain response*

Q HOW MUCH DO I NEED TO MEDITATE TO REDUCE PAIN?

Research has shown that less than 1½ hours of meditation training may help alleviate pain and diminish pain-related brain changes. One study showed that just four 20-minute mindfulness classes reduced the unpleasantness of pain by 57 per cent and the intensity of pain by 40 per cent. It wasn't just the perception of the pain that changed: the brain's activity also measurably changed. The same study showed, via fMRI scans, that meditation reduced pain-related activation of the primary somatosensory cortex. Instead of a spike of activity in the area of the somatosensory cortex related to the location of the pain, researchers found that, while meditating, participants had more brain activity reflecting sensory awareness of the neck and throat, which represented the participants' mindfulness of their breathing.

Did you know?

CHRONIC PAIN CAUSES **GREY MATTER DETERIORATION**, BUT THE AREAS OF THE BRAIN THAT ARE DEGRADED BY CHRONIC PAIN ARE **RESTORED DURING MEDITATION** THROUGH INCREASING NEURAL CONNECTIONS IN THOSE AREAS.

YOGA THERAPY

Yoga therapy is a growing field in integrative healthcare, based on the mounting research into yoga's therapeutic benefit. With educational standards and a scope of practice beyond those of yoga teaching, yoga therapists use the tools of yoga to empower individuals towards wellbeing.

*Lifestyle changes and **mindset** shifts from yoga can help people **move beyond** a disease focus to cultivate **human flourishing***

Q WHAT CAN I EXPECT FROM A YOGA THERAPY SESSION?

Yoga therapy sessions are often one-on-one or in small groups of people with similar conditions or life situations. Yoga therapists will always take your health history into account and, though they don't make medical diagnoses, they provide an individualized assessment of your health using tools including:

- **observations of posture**, movement, and breath
- **questions about** mood and lifestyle
- **observations through the lens** of yogic subtle anatomy, such as the vayus and the five koshas.

The koshas are five layers, or "sheaths", that make up your self, similar to the layers of an onion. The koshas start with your physical wellbeing and end with bliss (see below). Yoga therapists consider all aspects of your wellbeing and how they interact in their recommendations. For example, arthritis in your physical body may be affecting your emotions and deeper connection to bliss, while your emotions may be exacerbating the pain. From these observations and considerations, yoga therapists create a personalized plan of care for each client using tools such as poses, breathwork, meditations, and lifestyle suggestions.

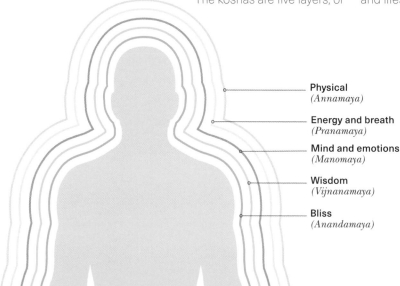

Physical
(Annamaya)

Energy and breath
(Pranamaya)

Mind and emotions
(Manomaya)

Wisdom
(Vijnanamaya)

Bliss
(Anandamaya)

THE FIVE KOSHAS
Each of these five layers or "sheaths" must be looked after if we are to live a healthy, balanced life.

Q HOW DOES YOGA THERAPY WORK?

Yoga has profound therapeutic potential because it acts on what researchers call a biopsychosocial-spiritual model (see right). Much of yoga research is done through this lens, showing therapeutic yoga's promise for multidimensional conditions such as chronic pain, trauma, and anxiety. Just as with the koshas (see left), the core of yoga therapy is that each aspect of self interacts with the others. To address this, yoga therapy applies a balance of research evidence, client values, and clinician experience.

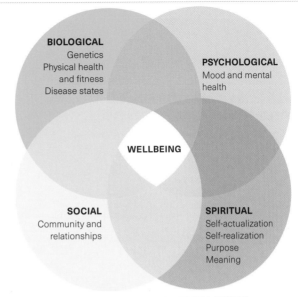

BIOLOGICAL
Genetics
Physical health and fitness
Disease states

PSYCHOLOGICAL
Mood and mental health

WELLBEING

SOCIAL
Community and relationships

SPIRITUAL
Self-actualization
Self-realization
Purpose
Meaning

BIOPSYCHOSOCIAL-SPIRITUAL MODEL

Q DO WE HAVE SCIENTIFIC EVIDENCE TO SUPPORT THE BENEFITS OF YOGA THERAPY?

Yes. The vast majority of scientific research into yoga is focused on understanding its therapeutic benefits, particularly for one of the world's most pressing healthcare issues: lifestyle-based chronic diseases (see pp.178–79, pp.188–91, and pp.194–95). The quality of this research is also improving, although some of the therapeutic benefits of yoga may never be fully understood through Western scientific inquiry. The yoga therapy profession is now growing partly because of the need for highly trained individuals who can work with specialized populations, for example veterans and those in cancer care.

Q HOW DOES YOGA THERAPY COMPARE TO OTHER HEALTHCARE PRACTICES?

Most healthcare systems work on the level of pathogenesis, which is a disease-based model of healthcare. The primary aim in this model is managing symptoms and "fixing" parts and pieces of the system. Although yoga therapy often is successful at managing symptoms, such as by providing pain relief, it also works on a level of salutogenesis, which is a health-based model. Rather than focusing on the disease to be cured or a problem to be fixed, salutogenesis focuses on creating wellbeing. Lifestyle changes and mindset shifts from yoga, therefore, can help people move beyond a disease focus to cultivate human flourishing.

TRANSFORMATION

Exercise is the most common reason why people first come to yoga. However, the spiritual side of yoga often becomes more important for those who continue to practise. With advances in technology including neuroimaging, researchers are now exploring yoga's potentially transformative spiritual effects.

*Neuroscientists are now studying the **brain** during **spiritual** states*

Q WHAT ARE THE SPIRITUAL STATES THAT ANCIENT YOGIS SPOKE OF?

The "eight limbs" of yoga are outlined in an ancient text called the "Yoga Sutras". The first four limbs concern how we live in the external world, and are intended to prepare your body and mind for the second four, which concern our internal world or consciousness (see below).

Astronauts undergo a similar process to the eight limbs of yoga: from an ethical code to intensive physical exercises to prepare the body and mind. When in space, "Earth gazing" is reportedly so captivating that astronauts spend hours just staring at the planet. This can be seen as similar to yogic concentration (dharana) exercises, such as staring at the flame of a candle to improve concentration and eventually evoke higher states of consciousness.

According to a 2016 paper called "The Overview Effect: Awe and Self-Transcendent Experience in Space Flight", astronauts return to Earth with a new perspective and sense of purpose. The founder of Phoenix Rising Yoga Therapy, Michael Lee, believes we can experience the same transformation on Earth by exploring the last four limbs of yoga.

3 POSES AS EXERCISE (ASANAS)

4 BREATHWORK (PRANAYAMA)

5 CONTROLLED SENSES (PRATYAHARA)

2 SELF-REGULATION (NIYAMAS)

YOGA

6 CONCENTRATION (DHARANA)

1 SELF-CONTROL (YAMAS)

8 ENLIGHTENMENT (SAMADHI)

7 MEDITATION (DHYANA)

THE EIGHT LIMBS OF YOGA
The ultimate aim of the eight limbs of yoga is to help us live a meaningful life. Not all modern classes incorporate them, but many at least allude to this depth and potential.

HOW CAN WE STUDY THE EFFECTS OF SPIRITUALITY?

Neuroscientists are now studying the brain during spiritual states, with fascinating findings. American neuroscientist Dr Andrew Newberg from the Marcus Institute of Integrative Health, for example, uses neuroimaging to understand higher spiritual states, including spiritually based meditative states such as Samadhi, deep prayer practices, and some drug-induced spiritual experiences (see below).

Four common brain patterns in spiritual experiences

Dr Newberg has compared the brain at rest and during transcendent spiritual experiences, including Samadhi, to identify specific brain activity patterns associated with spirituality.

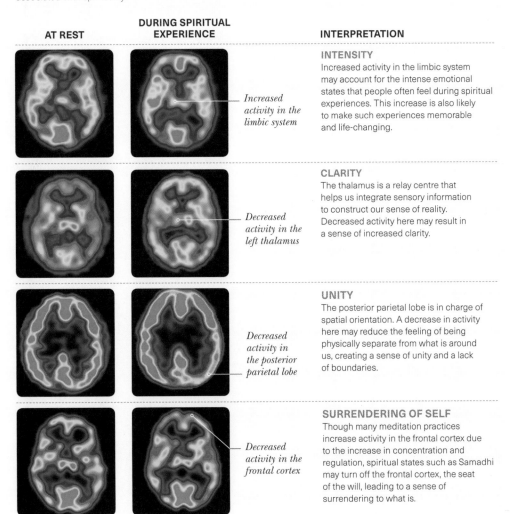

AT REST	DURING SPIRITUAL EXPERIENCE		INTERPRETATION
		Increased activity in the limbic system	**INTENSITY** Increased activity in the limbic system may account for the intense emotional states that people often feel during spiritual experiences. This increase is also likely to make such experiences memorable and life-changing.
		Decreased activity in the left thalamus	**CLARITY** The thalamus is a relay centre that helps us integrate sensory information to construct our sense of reality. Decreased activity here may result in a sense of increased clarity.
		Decreased activity in the posterior parietal lobe	**UNITY** The posterior parietal lobe is in charge of spatial orientation. A decrease in activity here may reduce the feeling of being physically separate from what is around us, creating a sense of unity and a lack of boundaries.
		Decreased activity in the frontal cortex	**SURRENDERING OF SELF** Though many meditation practices increase activity in the frontal cortex due to the increase in concentration and regulation, spiritual states such as Samadhi may turn off the frontal cortex, the seat of the will, leading to a sense of surrendering to what is.

ON THE FRONTIERS OF SCIENCE

Scientists predict that we only observe and understand 4 per cent of the universe we live in. Similarly, we are only on the frontiers of exploration when it comes to the science of the human brain, mind, and consciousness, which gets to the heart of yoga's capacity for transformation.

*Bear in mind that **extraordinary claims** require **extraordinary evidence***

Q HOW DO I KNOW IF A YOGA STUDY IS RELIABLE?

Not all yoga research is created equally, so it is good to approach it critically. Some factors to consider are:

● **What kind of study is it?** The hierarchy of scientific evidence (see below) gives a good idea of how reliable different kinds of studies are. Evidence that is lower on the pyramid is still valuable, but the higher up, the more reliable. There are increasing systematic reviews and meta-analyses on key topics in yoga, including mental health, heart disease, chronic pain, and safety.

● **How large is the sample size?** From case reports of one to randomized controlled trials (RCTs) of 228 people, yoga studies tend to be relatively small, especially compared to pharmaceutical RCTs with up to tens of thousands of participants.

● **Is there a control group?** If so, what? Many yoga studies incorporate a "usual care" control group. A few higher quality ones have an active control, such as comparing yoga to exercise or talk therapy.

● **What is the conclusion?** Bear in mind that extraordinary claims require extraordinary evidence. This is why many yoga researchers use phrases such as "yoga may improve" or "this suggests that yoga helps". As interest in yoga research increases, scientists will keep questioning results.

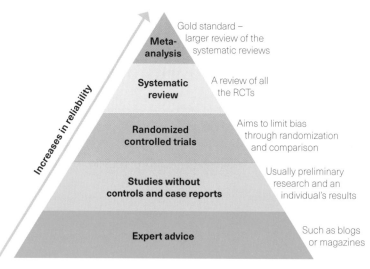

Increases in reliability →

Meta-analysis — Gold standard – larger review of the systematic reviews

Systematic review — A review of all the RCTs

Randomized controlled trials — Aims to limit bias through randomization and comparison

Studies without controls and case reports — Usually preliminary research and an individual's results

Expert advice — Such as blogs or magazines

HIERARCHY OF EVIDENCE
You can use this pyramid to gauge the reliability of different types of scientific evidence.

Q IS THERE SCIENTIFIC EVIDENCE TO SUPPORT YOGIC CONCEPTS, SUCH AS PRANA AND THE CHAKRAS?

Yoga research tends to focus on specific health conditions and practical benefits, rather than subtle energetics, as prana and chakras represent a way of knowing that doesn't necessarily translate directly to a straightforward analysis of biology. Some people, for example, claim that the flow of prana is in alignment with the nerves, and the chakras with the glands, but there is no scientific evidence to support this. It may be that before dissection showed us where these structures were, yogis felt them working in their bodies. It is also possible that we are still limited by our current instruments and will one day have the tools to locate and measure prana.

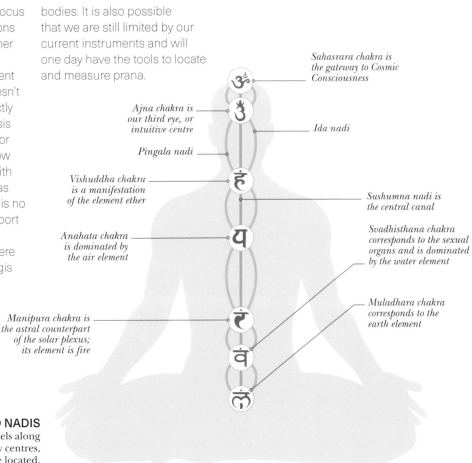

Sahasrara chakra is the gateway to Cosmic Consciousness

Ajna chakra is our third eye, or intuitive centre

Ida nadi

Pingala nadi

Vishuddha chakra is a manifestation of the element ether

Sushumna nadi is the central canal

Svadhisthana chakra corresponds to the sexual organs and is dominated by the water element

Anahata chakra is dominated by the air element

Muladhara chakra corresponds to the earth element

Manipura chakra is the astral counterpart of the solar plexus; its element is fire

CHAKRAS AND NADIS
Nadis are energy channels along which the seven energy centres, known as chakras, are located.

Q HOW MUCH RESEARCH IS THERE INTO YOGA?

Research into yoga is relatively limited, albeit on the rise, and fast. A bibliometric review of the relevant research from 1967 to 2013, for example, showed an exponential growth in the number of studies conducted from fewer than 25 publications from 1967–1973 to well over 225 publications in 2009–2013, correlating with the rise of popularity of yoga.

The research also identifies that the top four areas of research are:

- mental health disorders
- cardiovascular disease
- respiratory diseases
- musculoskeletal disorders.

CAUTIONS

Just as the Hippocratic Oath states "first do no harm", the first principle of yoga is *ahimsa*, which translates to non-harm. To avoid harm, it is important to know your body and adapt or modify poses and practices based on your needs and health conditions. Everyone is different, so use these pages as a general guide.

Injuries in yoga do happen, as they do in all types of physical activity, from walking down the stairs to lifting weights at the gym. A meta-analysis of randomized controlled trials, however, found that yoga is as safe as other types of recommended exercise. In fact, yoga may be safer than many forms of exercise because it often incorporates slow transitions, present moment awareness, and an emphasis on non-harm.

That said, if you believe that yoga practices are powerful enough to profoundly benefit you, you must also acknowledge that yoga has the power to harm, and you must treat it with that level of respect. To prevent injury, therefore, practise the first two limbs of yoga – the Yamas and Niyamas – both in yoga class and in life (see p.205). It is also advisable to bear in mind the following guidelines:

- **We all have** differently shaped bones and bodies, so poses will look different when practised by different people. Some postures may not be accessible to you without modifications
- **Allow recovery** after strains, sprains, tears, breaks/fractures, surgery, or wounds. After surgery, ask your surgeon for guidance

- **The point of yoga** is not to be able to perform an asana perfectly, or to do any particular technique or pose. Enjoy the journey of self-exploration!
- **Avoid anything that** causes pain or increases existing pain
- **Be cautious of** sharp sensations inside the body or sharp, shooting sensation down the limbs
- **Avoid anything that causes numbness** in the limbs.

CONDITIONS

The following pages outline any cautions and considerations for specific health conditions that you should bear in mind when practising yoga, as general guidance. However, you should always ask your professional medical team what is right for you. If in doubt, work with a qualified yoga professional, such as a yoga therapist.

ACID REFLUX/GERD/HEARTBURN
Be cautious of or avoid any full or partial inversion where the head goes below the heart, and fast breathing (kapalabhati).

ANKYLOSING SPONDYLITIS
Be cautious of spinal flexion and move slowly into gentle spinal extension.

ANXIETY/TENDENCY TOWARDS PANIC ATTACKS
Be cautious of inversions, backbends, fast breathing (kapalabhati), or breath holding (kumbhaka) during symptoms.

ARTHRITIS (including osteoarthritis and rheumatoid arthritis, and other conditions that involve joint inflammation)
For osteoarthritis and rheumatoid arthritis, avoid anything that increases joint pain, and focus on modifying poses for comfort, strengthening, and learning to meditate to cope with pain; for rheumatoid arthritis, avoid hot yoga and overheating.

ASTHMA
Be cautious when practising backbends, holding the breath (kumbhaka), and fast breathing (kapalabhati); avoid intense back bending during symptoms.

BURSITIS AND TENDONITIS
Avoid anything that increases pain or swelling; rest the affected area during acute stages.

CARPAL TUNNEL SYNDROME
Be cautious of or avoid arm balances or weight bearing while wrists are extended (e.g. Plank or

Crow Pose), especially if it increases numbness; consider resting your forearms on the floor or blocks, or try using a wedge.

DEGENERATIVE DISC DISEASE

Practise spinal flexion and spinal rotation gently; be cautious during or avoid headstands, shoulderstands, or anything that puts pressure on the neck.

DIABETES

For Type 1, avoid anything that puts pressure on your insulin pump; for Type 1 and 2, eat before class if needed, and rest if lightheaded.

DISC HERNIATION (SLIPPED, BULGING, PROTRUDING)

Be cautious of unsupported spinal flexion, such as a Standing or Seated Forward Fold, and spinal rotation; focus on lengthening the spine before gently entering a pose, and consider keeping the spine neutral and bending at the hips into a Forward Fold – Child's or Cat Pose may be safer forms of spinal flexion; be cautious during headstands, shoulderstands, or anything that puts pressure on the neck.

EAR INFECTION

Be cautious with inversions and in balancing poses.

EYE CONDITIONS THAT INCREASE PRESSURE (such as glaucoma, detached retina, diabetic retinopathy, or recent cataract surgery)

Be cautious with or avoid any pose in which the head goes below the heart, breath holding (kumbhaka), and fast breathing (kapalabhati); seek the advice of your ophthalmologist if you are unsure.

FIBROMYALGIA

Consider restorative yoga and yoga nidra; use lots of props and let your teacher know if you prefer not to be touched in a hands-on assist.

FROZEN SHOULDER (ADHESIVE CAPSULITIS)

Move slowly into shoulder stretches and gradually increase the stretch over time.

HEART CONDITIONS

Be cautious when performing inversions, breath holding (kumbhaka), and fast breathing (kapalabhati); you should also seek the advice of your cardiologist.

HIGH BLOOD PRESSURE (HYPERTENSION)

Be cautious with any pose where the head goes below the heart, breath holding (kumbhaka), and fast breathing (kapalabhati); if your blood pressure is not currently regulated, avoid full inversions, intense practice, and hot yoga completely.

HIP REPLACEMENT

Follow these precautions 6–8 weeks after surgery, and with the advice of your doctor. In the anterior approach, be cautious of or avoid extension (as in the lifted leg in Warrior III); in the posterior approach, be cautious of or avoid hip flexion past 90°, internal rotation, and crossing the midline (crossing legs); after proper healing, you are likely to be able to perform all of these movements, but move slowly into the poses and ask your doctor for advice.

HYPERMOBILITY

Avoid any extreme movement or hyperextension of joints; focus on strengthening.

KNEE LIGAMENT INJURY (ACL, PCL, LCL, MCL)

Be cautious with poses that involve rotation (e.g. Triangle Pose and Warrior II); for ACL, avoid deep knee flexion and for PCL be cautious of hyperextension/locking your knees; for both, be cautious of or avoid jumping into poses.

Continued →

CAUTIONS *continued*

KNEE MENISCUS TEAR/INJURY
Be cautious of or avoid deep knee flexion, especially if weight bearing.

KNEE REPLACEMENT
Avoid extreme knee flexion; cushion the knee with blankets or padding when in kneeling poses.

LOW BLOOD PRESSURE (HYPOTENSION)
Move slowly out of any pose where the head goes below the heart; pause a few moments in a restful pose, such as Child's Pose, after full inversions to prevent dizziness; move slowly when rising from the floor.

MIGRAINE
Be cautious when performing full inversions; try practising in a room with dimmed light.

MULTIPLE SCLEROSIS
Be cautious of intense practices that make you feel overheated; avoid hot yoga.

OBESITY
Be cautious of unsupported spinal flexion and full inversions, such as headstands, shoulderstands, or anything that puts your weight on your neck.

OSTEOPOROSIS/OSTEOPENIA
For spinal areas, talk to your doctor as what you can do will depend on the severity of your condition.

However, general guidelines are to be cautious of unsupported spinal flexion and spinal rotation; move slowly and focus on elongating the spine before coming into twists, and consider flexing at the hips and try keeping the spine neutral in many Forward Folds to avoid the risks of spinal flexion (Child's or Cat Pose may be safer forms of spinal flexion); take extreme caution during or avoid headstands, shoulderstands, or anything that puts pressure on the neck; take particular caution to move slowly and gently in movements that combine spinal flexion and rotation such as Triangle Pose; take care in transitioning poses and balancing poses to reduce the risk of falling; for non-spinal areas, such as hips or wrists, move slowly into poses and focus on mindfully strengthening muscles around the affected areas.

PARKINSON'S DISEASE
Be cautious of inversions and balancing poses; try holding onto the wall or a chair to prevent falls; use props as needed.

PLANTAR FASCIITIS
Take caution with or avoid jumping into poses, or any movement that exacerbates symptoms; stretch the feet and legs slowly and mindfully.

PREGNANCY
Be cautious of full inversions, especially if you don't already have an inversion practice; be cautious of or avoid anything that puts pressure on the abdomen (e.g. Locust Pose or extreme abdominal engagement); avoid extreme abdominal stretching (e.g. Wheel Pose); don't stay too long lying on your back in later stages of pregnancy if uncomfortable, and consider lying on your side with a pillow between your legs, or propping yourself up to lie at an angle.

ROTATOR CUFF (TEAR, TENDONITIS, INSTABILITY)
Be cautious with any shoulder stretches; avoid Low Plank Pose (Chaturanga), particularly in acute stages; focus on strengthening over stretching, e.g. consider holding a forearm version of Plank or Downward Dog on the floor or wall.

SACROILIAC (SI) DYSFUNCTION/PAIN
Avoid extreme twists and be cautious in widelegged postures (e.g. Triangle Pose); be cautious that being in asymmetric poses, such as Warrior poses or Triangle Pose for a prolonged period on one side may be uncomfortable. If so, consider switching sides more often.

SCIATICA
Be cautious of anything that increases numbness; if the condition is due to a tight piriformis, consider modified versions of Pigeon Pose, e.g. Figure 4 on your back.

SCOLIOSIS

Avoid anything that causes numbness; consider strengthening your back muscles by practising Side Plank Pose and gently stretching in the opposite direction of the curvature.

SHOULDER DISLOCATION, HISTORY OF

Avoid any extreme shoulder extension, especially while weight bearing, such as in Wheel Pose; consider focusing your practice on strengthening.

SINUSITIS

Be cautious of inversions and spinal extension; you may find the alternate nostril breathing technique difficult.

SPINAL STENOSIS

Be cautious of spinal extension.

SPONDYLOLISTHESIS

Ask your doctor for what to avoid in your individual case; however, general guidance is: be cautious of spinal extension and spinal rotation; avoid deep twisting, moderate or deep backbends, and jumping into poses.

STROKE, HISTORY OR RISK OF

Be wary of inversions and extreme cervical extension; avoid anything that puts pressure on the neck.

VERTIGO/DIZZINESS

See Low Blood Pressure.

Approaching yoga with respect

The Yamas and Niyamas are the ethical guidelines for a yogic lifestyle. Traditionally, a guru would require that a practitioner lives these principles before learning any asana, to prevent ego and injury.

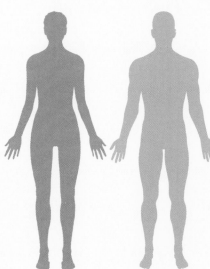

YAMAS (SELF-CONTROL)

- **Ahimsa (non-harm):** don't do anything that hurts or increases current pain
- **Satya (truthfulness):** be truthful with yourself about what your body can do today
- **Asteya (non-stealing/abundance):** focus on the things you can do instead of what you cannot do
- **Brahmacharya (moderation):** practise everything in moderation to regulate your energy
- **Aparigraha (non-possession):** there is no need to grasp for a body you used to have, or to be jealous of the person practising next to you.

NIYAMAS (SELF-REGULATION)

- **Saucha (cleanliness):** organize your props and practice area to prevent falls or distraction
- **Santosha (contentment):** find contentment with where you are physically and mentally today
- **Tapas (self-discipline):** balance your burning desire to improve with the practice of non-harm
- **Svadhyaya (self-study):** observe your breath and energy today and adjust your practice to respect that
- **Ishvara Pranidhana (surrendering/accepting):** allow a sense of surrendering to what is in the present moment, changing what you can (for example, using a prop for comfort in a pose), but accepting what you cannot change. Just be.

GLOSSARY

Acute When symptoms come on rapidly; acute pain generally lasts for less than 3–6 months.

Alignment In yoga, the way a pose is instructed with the intention of encouraging optimal function and preventing injury; although there are general principles, proper alignment may vary from person to person and day to day, and based on the intention behind the pose.

Anatomy Study of the structure of the body, including the naming of parts.

Antigen Invader that the body's immune system fights with antibodies and white blood cells.

Arthritis Group of joint conditions that involves joint inflammation and/or damage; osteoarthritis is the most common type and involves damage to the cartilage of the joint due to wear and tear.

Asana Yoga pose or posture.

Bile Substance that helps break down fats in digestion.

Cartilage Firm but flexible connective tissue; includes hyaline (glass-like, in synovial joints to reduce friction), fibrocartilage (firm cushioning, in intervertebral discs for cushioning), and elastic (stretchy, in nose and ears for elasticity).

Central nervous system (CNS) The brain and spinal cord; controls the body and perceives the world.

Cerebral cortex Outer shell of the cerebrum.

Cerebrum Largest part of the brain; contains the cerebral cortex and some internal structures such as the hippocampus.

Cervical spine Seven vertebrae of the neck.

Chromosomes Thread-like molecules made of DNA and proteins; humans generally have 23 pairs.

Chronic Long-lasting symptoms, disease, and/or pain; chronic pain generally persists for longer than 3–6 months.

Collagen Key component in many connective tissues; has good tensile strength, allowing it to resist tension or pull.

Concentric contraction Muscle shortening in response to a load, as in lifting a weight in a biceps curl.

Connective tissue Forms connections in your body; subtypes include cartilage, bone, blood, lymph, adipose (fat), and elastic tissue (such as in the ears and nose), along with fibrous connective tissue.

Control group The research group that doesn't receive the intervention being studied; may recieve nothing, or an active control, to act as a comparison.

Deep Further inward from the surface; for instance, your muscles are deep to your skin.

Diaphragm Usually refers to the respiratory diaphragm, which is the primary muscle used in a relaxed breath; there are also the vocal/thoracic outlet diaphragm and urogenital/pelvic floor diaphragm.

DNA Deoxyribonucleic acid; carries hereditary information in genes; within chromosomes.

Eccentric contraction Muscle lengthening in response to a load, as in lowering a weight in a biceps curl.

Engaging When a muscle is contracting; "Engaging while stretching" is used in this book to describe contraction while a muscle is in a neutral or lengthening position, as in an eccentric contraction, but held steady.

Epithelial tissue Forms coverings in your body, such as the superficial layer of skin.

Fascia Fibrous connective tissue that surrounds muscles and other organs.

Fibrous connective tissue Contains either a parallel or irregular pattern of collagen fibres; includes dense regular connective tissue, of tendons and ligaments, and dense irregular connective tissue, of fascia and synovial joint capsules.

fMRI Functional magnetic resonance imaging; machine that measures blood flow in the brain to reflect neural activity.

Grey matter Tissue in the central nervous system that contains mostly cell bodies, dendrites, and synapses (as compared to white matter which contains mostly axons and is white due to myelin).

Heart rate variability (HRV) Measure of the variation between heart beats within a specific increment of time; may be an indicator of cardiorespiratory and stress resilience.

Hip points Colloquial name for the two bony protrusions on the front of the pelvis, called the anterior superior iliac spines.

Homeostasis State of dynamic equilibrium maintained in the human body to support life.

Hot yoga Yoga classes in rooms heated to 33–40.5°C (92–105°F).

Hyperextension Extreme extension of a joint, often past normal range.

Hypermobile Extremely flexible beyond normal limits.

Hypertension High blood pressure.

Inflammation Indication that the body is fighting something locally or systemically; symptoms can include redness, swelling, heat, and pain.

Interoception Sensory body awareness of your internal environment, including of the digestive organs, heart, and muscles.

Intervertebral disc Discs, made mostly of fibrocartilage, which absorb shock in between vertebrae and allow movement.

Inversion Poses, like Headstand, where the body is "upside-down"; partial inversions include any pose where the head is below the heart.

Isometric contraction Muscle engagement where the muscle stays the same length, such as pushing into a wall or the floor.

Isotonic contraction Muscle engagement where the muscle changes length; can either be eccentric or concentric.

Kinesiology Study of body movement..

Kumbhaka Pranayama practice of breath retention.

Kyphosis Convex curves of the spine, found naturally in the thoracic spine and sacrum; the term can also describe an excessive amount of this convex curve, as in a Dowager's hump.

Ligament Connects bone to bone; made of dense regular connective tissue proper, which has parallel collagen fibres.

Lordosis Concave curves of the spine, found naturally in the lumbar and cervical spine; the term can also describe an excessive amount of this concave curve.

Lumbar spine Five vertebrae of the low back.

Lymph Fluid filled with white blood cells to fight invaders; collected from interstitial fluid, it drains back into the heart after being filtered in lymph nodes.

Meditation Concentration or mental focus exercise; includes mindfulness, mantra, loving-kindness, transcendental meditation (TM), and others; Dhyana, in Sanskrit.

Meta-analysis Systematic assessment of previous research in a specific area to derive broad conclusions; the gold standard of review articles.

Mindfulness Paying attention on purpose to the present moment, without judgement (as defined by researcher Jon Kabat Zinn, PhD).

Muscle tissue Contractile tissue; the three types are skeletal, smooth, and cardiac muscle.

Nadis According to Indian medicine and Hindu philosophy, these are channels for prana to flow.

Nerve Bundle of axons of neurons in the peripheral nervous system; conductive tissue that acts like wires through the body, carrying signals to and from the central nervous system. Includes cranial nerves and spinal nerves; a bundle of axons in the central nervous system is called a tract.

Nervous tissue Conductive tissue made of neurons and helper cells.

Neuron Nerve cell; carries electrical signals.

Neuroplasticity Ability of the brain to create neural connections.

Neutral spine Position of optimal load distribution for the spine; maintains the natural curves of the cervical (lordosis), thoracic (kyphosis), and lumbar (lordosis) segments of the spine.

Neutral pelvis Position of the pelvis that best supports the inward curve of the lumbar spine. No excessive anterior or posterior pelvic tilt; hip points are in line with each other; minimized stress on ligaments, muscles, and other tissues.

Osteoporosis Condition where bones become weak and brittle, leaving them at higher risk for fractures.

Parasympathetic nervous system (PSNS) "Rest and digest" branch of the autonomic nervous system; the relaxation response.

Peripheral nervous system (PNS) Includes the cranial and spinal nerves.

Physiology The study of the function of parts and systems in the body; the study of how the body works.

Postural hypotension Also called orthostatic hypotension; a sudden onset of low blood pressure caused by standing up too quickly from the floor or an inversion.

Prana Sanskrit word meaning life-force energy, vital energy, or breath, similar to the Chinese concept of qi; yogis believe you can consciously transform and move your prana.

Pranayama Sanskrit word meaning breath extension or control; breathwork.

Proprioception Type of interoception that focuses on spatial body awareness, particularly while in motion.

Randomized controlled trial (RCT) Randomization of the experimental group and control(s), which can lead to less bias; gold standard of research trials.

Sacroiliac joint Joint between the sacrum and ilium of the pelvis; allows a small amount of movement.

Samskaras According to Indian philosophy, these are imprints or impressions of our past actions.

Sanskrit The ancient Indian language that many yoga texts were written in.

Stretching When muscle fibres lengthen, often beyond resting length.

Sun salutation Series of asanas done in flowing sequence to warm up the body and focus the mind.

Superficial Closer to the surface; for instance, your skin is superficial to your muscles.

Supine Lying on your back, face upwards.

Sympathetic nervous system (SNS) "Fight or flight" branch of the autonomic nervous system; the stress response.

Synovial joint Most common and most mobile type of joint in the body, such as the shoulders, hips, and knees.

Tendon Connects muscle to bone; made of dense regular connective tissue proper, which has parallel collagen fibres.

Thoracic spine The 12 vertebrae of the mid-back region.

Tissues Collection of cells that come together for a similar function; the four primary tissue types are epithelial, connective, muscle, and nervous.

Vagus nerve Tenth cranial nerve (CN X), important in parasympathetic control of the heart, lungs, and digestive organs.

Vayus According to yoga philosophy, your prana flows in specific patterns called the vayus: Prana (in), Udana (into head), Vyana (into limbs), Samana (around), and Apana vayu (down and out).

Yoga therapy According to the International Association of Yoga Therapists, "Yoga therapy is the process of empowering individuals to progress towards improved health and well-being through the application of the teachings and practices of Yoga"; this developing field has educational standards that exceed those for general yoga instruction, and prepare practitioners to work safely with health conditions.

INDEX

BIBLIOGRAPHY

10–11 R. Chaix et al., "Epigenetic clock analysis in long-term meditators", *Psychoneuroendocrino* 85 (2017); E. Epel et al., "Can Meditation Slow Rate of Cellular Aging? Cognitive Stress, Mindfulness, and Telomeres", *Ann NY Acad Sci* 1172 (2009); D. Ornish et al., "Effect of comprehensive lifestyle changes on telomerase activity and telomere length in men with biopsy-proven low-risk prostate cancer: 5-year follow-up of a descriptive pilot study", *Lancet Oncol* 14 (2013). **12–17** S. H. Moonaz et al., "Yoga in Sedentary Adults with Arthritis: Effects of a Randomized Controlled Pragmatic Trial", *J Rheumatol* 42 (2015); S. Muraki et al., "Quadriceps muscle strength, radiographic knee osteoarthritis and knee pain: the ROAD study", *BMC Musculoskel Dis* 16 (2015); M. Wallden, "The neutral spine principle", *J Bodywork Movement Ther* 13 (2009). **18–21** T. W. Myers, *Anatomy trains* (3rd ed.), Edinburgh, Churchill Livingstone/Elsevier, 2014. **22–27** M. Balasubramaniam et al., "Yoga on our minds: a systematic review of yoga for neuropsychiatric disorders", *Front Psychiat* 3 (2013); B. Rael Cahn et al., "Yoga, Meditation and Mind-Body Health: Increased BDNF, Cortisol Awakening Response, and Altered Inflammatory Marker Expression after a 3 Month Yoga and Meditation Retreat", *Front Hum Neurosci* 11 (2017); R. A. Gotink et al., "Meditation and yoga practice are associated with smaller right amygdala volume: the Rotterdam study", *Brain Imaging Behav* (2018); B. K. Hölzel et al., "Mindfulness practice leads to increases in regional brain gray matter density", *Psychiat Res Neuroim* 191 (2011); D. E. Larson-Meyer, "A Systematic Review of the Energy Cost and Metabolic Intensity of Yoga", *Med Sci Sport Exer* 48 (2016). **28–29** M. Á. D. Danucalov et al., "Cardiorespiratory and Metabolic Changes during Yoga Sessions: The Effects of Respiratory Exercises and Meditation Practices", *Appl Psychophys Biof* 33 (2008); K. E. Innes and T. K. Selfe, "Yoga for Adults with Type 2 Diabetes: A Systematic Review of Controlled Trials", *J Diabetes Res* 2016 (2016); C. C. Streeter et al., "Effects of yoga on the autonomic nervous system, gamma-aminobutyric-acid, and allostasis in epilepsy, depression, and post-traumatic stress disorder", *Med Hypotheses* 78 (2012). **30–33** S. Telles et al., "Blood Pressure and Heart Rate Variability during Yoga-Based Alternate Nostril Breathing Practice and Breath Awareness", *Med Sci Monitor Basic Res* 20 (2014); M. Joshi and S. Telles, "Immediate effects of right and left nostril breathing on verbal and spatial scores", *Indian J Physiol Pharmacol* 52 (2008); R. Kahana-Zweig et al., "Measuring and Characterizing the Human Nasal Cycle", *PLoS ONE* 11 (2016); M. Kuppusamy et al., "Effects of Bhramari Pranayama on health – A systematic review", *J Trad Complem Med* 8 (2018); D. S. Shannahoff-Khalsa et al., "Ultradian rhythms of autonomic, cardiovascular, and neuroendocrine systems are related in humans", *Am J Physiol* 270 (1996); A. N. Sinha et al., "Assessment of the effects of pranayama/alternate nostril breathing on the parasympathetic nervous system in young adults", *J Clin Diag Res* 7 (2013); G. Yadav and P. K. Mutha, "Deep Breathing Practice Facilitates Retention of Newly Learned

Motor Skills", *Sci Rep* 6 (2016); F. Yasuma and J. Hayano, "Respiratory Sinus Arrhythmia", CHEST 125 (2004). **34–35** World Health Organization, "Cardiovascular diseases (CVDs)", *World Health Organization* [web article], 17 May 2017, (accessed 20 Aug 2018); H. Cramer et al., "Effects of yoga on cardiovascular disease risk factors: A systematic review and meta-analysis", *Int J Cardiol* 173 (2014); K. E. Innes et al., "Risk Indices Associated with the Insulin Resistance Syndrome, Cardiovascular Disease, and Possible Protection with Yoga: A Systematic Review", *J Am Board Fam Med* 18 (2005); D. Ornish et al., "Can lifestyle changes reverse coronary heart disease? The Lifestyle Heart Trial", *Lancet* 336 (1990). **36–37** Harvard Health Letter, "Inflammation: A unifying theory of disease", *Harvard Health Publishing* [web article], Apr 2006, (accessed 20 Aug 2018); R. I. Falkenberg et al., "Yoga and immune system functioning: a systematic review of randomized controlled trials", *J Behav Med* 41 (2018); T. Oka et al., "Changes in fatigue, autonomic functions, and blood biomarkers due to sitting isometric yoga in patients with chronic fatigue syndrome", *BioPsychoSocial Med* 12 (2018). **38–39** M. Berners-Lee et al., "The relative greenhouse gas impacts of realistic dietary choices", Energy Policy 43 (2012); H. C. J. Godfray et al., "Food Security: The Challenge of Feeding 9 Billion People", *Science* 327 (2010); M. Springmann et al., "Analysis and valuation of the health and climate change cobenefits of dietary change", *P Natl A Sci* 113 (2016); D. Tilman and M. Clark, "Global diets link environmental sustainability and human health", *Nature* 515 (2014). **40–41** S. Prosko, "Optimizing Pelvic Floor Health Through Yoga Therapy", *Yoga Ther Today*, 12 (2016); A. Huang et al., "PD32-01 A Randomized Trial of a Group-Based Therapeutic Yoga Program for Ambulatory Women With Urinary Incontinence", *J Urology* 199 (2018). **46–49** T. W. Myers, *Anatomy Trains* (3rd ed.), Edinburgh, Churchill Livingstone/Elsevier, 2014. **50–53** F. Dehghan et al., "The effect of relaxin on the musculoskeletal system", *Scand J Med Sci Spor* 24 (2013). **60–63** J. M. M. Brown et al., "Muscles within muscles: Coordination of 19 muscle segments within three shoulder muscles during isometric motor tasks", *J Electromyogr Kines* 17 (2007); H. Mason, "Learning to Abide with What Is: The Science of Holding Poses", *Yoga Ther Today* 13 (2017). **76–79** E. J. Benjamin et al., "Heart Disease and Stroke Statistics—2018 Update: A Report From the American Heart Association", *Circulation* 137 (2018); P. Page et al., *Assessment and Treatment of Muscle Imbalance: The Janda Approach*, Champaign (IL), Human Kinetics, 2010; K. W. Park et al., "Vertebral Artery Dissection: Natural History, Clinical Features and Therapeutic Considerations", *J Korean Neurosurg S* 44 (2008). **94–97** L. B. De Brito et al., "Ability to sit and rise from the floor as a predictor of all-cause mortality", *Eur J Prev Cardiol* 21 (2014); A. B. Newman et al., "Strength, but not muscle mass, is associated with mortality in the health, aging and body composition study cohort", *J Gerontol A-Biol* 61 (2006). **102–105** J. L. Oschman et al., "The effects of grounding (earthing) on inflammation, the immune response, wound healing, and prevention and treatment of chronic inflammatory and autoimmune diseases", *J Inflamm Res* 2015 (2015). **118–121** Y. H. Lu et al., "Twelve-Minute Daily Yoga Regimen Reverses Osteoporotic Bone Loss", *Top Geriatr Rehabil* 32 (2016). **128–131** L. M. Fishman et al., "Yoga-Based Maneuver Effectively Treats Rotator Cuff Syndrome", *Top Geriatr Rehabil* 27 (2011); R. Hector and J. L. Jensen, "Sirsasana (headstand) technique alters head/neck loading: Considerations for safety", *J Bodywork Movement Ther* 19 (2015); T. B. McCall, *Yoga as Medicine: The Yogic Prescription for Health and Healing*, New York, Bantam, 2007. **132–135** M. Robin, *A 21st-Century Yogasanalia: Celebrating the Integration of Yoga, Science, and Medicine*, Tucson (AZ), Wheatmark Inc., 2017. **136–139** P. Page et al., *Assessment and Treatment of Muscle Imbalance: The Janda Approach*, Champaign (IL), Human Kinetics, 2010. **146–149** L. B. De Brito et al., "Ability to sit and rise from the floor as a predictor of all-cause mortality", *Eur J Prev Cardiol* 21 (2014); R. T. Proyer, "The well-being of playful adults: Adult playfulness, subjective well-being, physical well-being, and the pursuit of enjoyable activities", *Eur J Humour Res* 1 (2013); United Nations, "Convention on the Rights of the Child", 2 Sep 1990, (accessed 11 Aug 2018). **150–153** D. Frownfelter and E. Dean, *Cardiovascular and Pulmonary Physical Therapy: Evidence to Practice* (4th ed.), St Louis, Elsevier Health Sciences, 2005. **154–157** L. M. Fishman et al., "Serial Case Reporting Yoga for Idiopathic and Degenerative Scoliosis", *Glob Adv Health Med* 3 (2014). **162–165** B. Duthey, "Background Paper 6.24 Low back pain", Priority Medicines for Europe and the World, World Health Organization, 2013; Society of Behavioral Medicine, "Yoga Shown to be Cost-Effective for Chronic Back Pain Management", *PR Web*, [web article], 13 Apr 2018, (accessed 17 Sep 2018). **166–169** H. Mason, "Learning to Abide with What Is: The Science of Holding Poses", *Yoga Ther Today* 13 (2017); W. D. Bandy and J. M. Irion, "The effect of time on static stretch on the flexibility of the hamstring muscles", *Phys Ther* 74 (1994). **170–173** J. Hamill and K. M. Knutzen, *Biomechanical Basis of Human Movement* (2nd ed.), Philadelphia, Wolters Kluwer Health, 2003. **176–177** K. deWeber et al., "Knuckle Cracking and Hand Osteoarthritis", *J Am Board Fam Med* 24 (2011); A. Guillot et al., "Does motor imagery enhance stretching and flexibility?", *J Sport Sci* 28 (2010); A. J. Hakim and R. Grahame, "A simple questionnaire to detect hypermobility: an adjunct to the assessment of patients with diffuse musculoskeletal pain", *Int J Clin Pract* 57 (2003); G. N. Kawchuk et al., "Real-Time Visualization of Joint Cavitation", *PLoS ONE* 10 (2015); V. K. Ranganathan et al., "From mental power to muscle power – gaining strength by using the mind", *Neuropsychologia* 42 (2004); D. Syx et al., "Hypermobility, the Ehlers-Danlos syndromes and chronic pain", *Clin Exp Rheumatol* 35 (2017). **178–179** R. Chaix et al., "Epigenetic clock analysis in long-term meditators", *Psychoneuroendocrinol* 85 (2017); L.-H. Chuang et al., "A Pragmatic Multicentered Randomized Controlled Trial of Yoga for Chronic Low Back Pain: Economic Evaluation", *Spine* 37 (2012); K. K. Hansraj, "Assessment of stresses in the cervical spine caused by posture and position of the head", *Surg Tech Int* 25 (2014); Society of Behavioral

Medicine, "Yoga Shown to be Cost-Effective for Chronic Back Pain Management", *PR Web*, [web article], 13 Apr 2018, (accessed 17 Sep 2018). **180–183** B. P. Acevedo et al., "The Neural Mechanisms of Meditative Practices: Novel Approaches for Healthy Aging", *Curr Behav Neurosci Reports* 3 (2016); R. F. Afonso et al., "Greater Cortical Thickness in Elderly Female Yoga Practitioners – A Cross-Sectional Study", *Front Aging Neurosci* 9 (2017); B. Bell and N. Zolotow, *Yoga for Healthy Aging: A Guide to Lifelong Well-Being*, Boulder, CO, Shambhala, 2017; A. J. Cerrillo-Urbina et al., "The effects of physical exercise in children with attention deficit hyperactivity disorder: a systematic review and meta-analysis of randomized control trials", *Child Care Hlth Dev* 41 (2015); B. Chethana et al., "Prenatal Yoga: Effects on Alleviation of Labor Pain and Birth Outcomes", *J Altern Complem Med* (2018); A. Herbert and A. Esparham, "Mind–Body Therapy for Children with Attention-Deficit/Hyperactivity Disorder", *Children* 4 (2017); Q. Jiang et al., "Effects of Yoga Intervention during Pregnancy: A Review for Current Status", *Am J Perinatol* 32 (2015); S. B. S. Khalsa and B. Butzer, "Yoga in school settings: a research review", *Ann NY Acad Sci* 1373 (2016); S. W. Lazar et al., "Meditation experience is associated with increased cortical thickness", *NeuroReport* 16 (2005); P. J. Reis and M. R. Alligood, "Prenatal Yoga in Late Pregnancy and Optimism, Power, and Well-Being", *Nurs Sci Quart* 27 (2014); M. Y. Wang et al., "Physical-Performance Outcomes and Biomechanical Correlates from the 32-Week Yoga Empowers Seniors Study", *Evid-Based Compl Alt* 2016 (2016). **184–185** B. K. Hölzel et al., "Mindfulness practice leads to increases in regional brain gray matter density", *Psychiat Res-Neuroim* 191 (2011); B. G. Kalyani et al., "Neurohemodynamic correlates of 'OM' chanting: A pilot functional magnetic resonance imaging study", *Int J Yoga* 4 (2011); K. Katahira et al., "EEG Correlates of the Flow State: A Combination of Increased Frontal Theta and Moderate Frontocentral Alpha Rhythm in the Mental Arithmetic Task", *Front Psychol* 9 (2018); F. Zeidan et al., "Mindfulness meditation improves cognition: Evidence of brief mental training", *Conscious Cogn* 19 (2010). **186–187** R. Anderson et al., "Using Yoga Nidra to Improve Stress in Psychiatric Nurses in a Pilot Study", *J Altern Complem Med* 23 (2017); H. Eastman-Mueller et al., "iRest yoga-nidra on the college campus: changes in stress, depression, worry, and mindfulness", *Int J Yoga Ther* 23 (2013); S. A. Gutman et al., "Comparative Effectiveness of Three Occupational Therapy Sleep Interventions: A Randomized Controlled Study", *OTJR-Occup Part Heal* 37 (2016); M. M. Hall et al., "Lactate: Friend or Foe", *Am Acad Phys Med Rehabil* 8 (2016); M. S. McCallie et al., "Progressive Muscle Relaxation", *J Hum Behav Soc Envir* 13 (2008); T. H. Nassif et al., "Mindfulness meditation and chronic pain management in Iraq and Afghanistan veterans with traumatic brain injury: A pilot study", *Milit Behav Heal* 4 (2016). **188–189** A. Ross et al., "National survey of yoga practitioners: Mental and physical health benefits", *Complement Ther Med* 21 (2013); M. B. Sullivan et al., "Yoga Therapy and Polyvagal Theory: The Convergence of Traditional Wisdom and Contemporary Neuroscience for Self-Regulation and Resilience", *Front Hum Neurosci* 12 (2018); S. Szabo et al., "'Stress' is 80 Years Old: From Hans Selye Original Paper in 1936 to Recent Advances in GI Ulceration", *Curr Pharm Des* 23 (2017); R. M. Yerkes and J. D. Dodson, "The relation of strength of stimulus to rapidity of habit-formation", *J Comp Neurol Psychol* 18 (1908). **192–193** R. A. Gotink et al., "Meditation and yoga practice are associated with smaller right amygdala volume: the Rotterdam study", *Brain Imaging Behav* (2018); P. A. Levine, *In an Unspoken Voice: How the Body Releases Trauma and Restores Goodness*, Berkeley (CA), North Atlantic Books, 2010; K. Nila et al., "Mindfulness-based stress reduction (MBSR) enhances distress tolerance and resilience through changes in mindfulness", *Ment Health Prev* 4 (2016); P. Payne et al., "Somatic experiencing: using interoception and proprioception as core elements of trauma therapy", *Front Psychol* 6 (2015); Y.-Y. Tang et al., "The neuroscience of mindfulness meditation", *Nat Rev Neurosci* 16 (2015). **194–195** M. C. Bushnell et al., "Cognitive and emotional control of pain and its disruption in chronic pain", *Nat Rev Neurosci* 14 (2015); E. J. Groessl et al., "Yoga for Military Veterans with Chronic Low Back Pain: A Randomized Clinical Trial", *Am J Prev Med* 53 (2017); G. L. Moseley and D. S. Butler, "Fifteen Years of Explaining Pain: The Past, Present, and Future", *J Pain* 16 (2015); N. Vallath, "Perspectives on Yoga inputs in the management of chronic pain", *Indian J Palliative Care* 16 (2010); F. Zeidan et al., "Mindfulness Meditation-Based Pain Relief Employs Different Neural Mechanisms Than Placebo and Sham Mindfulness Meditation-Induced Analgesia", *J Neurosci* 35 (2015); F. Zeidan et al., "The Effects of Brief Mindfulness Meditation Training on Experimentally Induced Pain", *J Pain* 11 (2010); F. Zeidan et al., "Brain Mechanisms Supporting Modulation of Pain by Mindfulness Meditation", *J Neurosci* 31 (2011). **196–197** International Association of Yoga Therapists, "Educational Standards for the Training of Yoga Therapists", *IAYT*, [web article], 2012, (accessed 10 Sep 2018); W. B. Jonas et al., "Salutogenesis: The Defining Concept for a New Healthcare System", *Global Adv Health Med* 3 (2014); International Association of Yoga Therapists, "Introduction to the IAYT Scope of Practice", *IAYT*, [web article], 2016, (accessed 10 Sep 2018); M. J. Taylor and T. McCall, "Implementation of Yoga Therapy into U.S. Healthcare Systems", *Int J Yoga Ther* 27 (2017). **198–199** C. L. Park et al., "Why practice yoga? Practitioners' motivations for adopting and maintaining yoga practice", *J Health Psychol* 21 (2014); M. T. Quilty et al., "Yoga in the Real World: Perceptions, Motivators, Barriers, and Patterns of Use", *Global Adv Health Med* 2 (2013); D. B. Yaden et al., "The overview effect: Awe and self-transcendent experience in space flight", *Psychol Consciousness* 3 (2016); A. B. Newberg, "The neuroscientific study of spiritual practices", *Front Psychol* 5 (2014). **200–201** M. Hagins and S. B. Khalsa, "Bridging yoga therapy and scientific research", *Int J Yoga Ther* 22 (2012); P. E. Jeter et al., "Yoga as a Therapeutic Intervention: A Bibliometric Analysis of Published Research Studies from 1967 to 2013", *J Altern Complem Med* 21 (2015). **202–203** H. Cramer et al., "The Safety of Yoga: A Systematic Review and Meta-Analysis of Randomized Controlled Trials", *Am J Epidemiol* 182 (2015).

Research on yoga is constantly evolving. For updated resources go to: **www.scienceof.yoga**

ABOUT THE AUTHOR

Ann Swanson, MS, C-IAYT, LMT, E-RYT500, is a mind-body science educator. She holds a Master of Science in yoga therapy from Maryland University of Integrative Health, where she continued on to become an adjunct faculty member. With years of experience tutoring and teaching anatomy and physiology in colleges, massage therapy schools, and yoga teacher training programmes, she has refined the ability to make complex scientific concepts simple to understand. Ann uniquely applies cutting-edge research practically to yoga while maintaining the heart of the tradition. In her private practice, she makes yoga therapy, qigong, and mindfulness meditation accessible and convenient in an online format, helping people worldwide manage pain and stress effectively.

For more about Ann, head to **www.AnnSwansonWellness.com**

ACKNOWLEDGMENTS

Author acknowledgments

Gratitude to my mentors and colleagues: Yogi Sivadas and Alice from Kailash Tribal School of Yoga in India, Yang Yan and Mahendra from Yogi Yogi in China, John Pace, Steffany Moonaz, Marlysa Sullivan, Laurie Hyland Robertson, and Michel Slover. Gratitude to my family: Mom, Dad, Joe, Aunt Sandy, and Pop. Gratitude to the brilliant DK team: Ruth, Clare, Arran, and everyone else.

Publisher acknowledgments

DK would like to thank Rebecca Fallowfield and Luca Frassinetti for production, Alison Gardner and India Wilson for design assistance, John Friend for proofreading, and Helen Peters for the index.

Picture credits